Botanical Research and Practices



# The Ethnobotanical Profile of Tehsil Kotli Sattian, Rawalpindi, Pakistan

# BOTANICAL RESEARCH AND PRACTICES

Additional books in this series can be found on Nova's website under the Series tab.

Additional e-books in this series can be found on Nova's website under the e-book tab.

BOTANICAL RESEARCH AND PRACTICES

# THE ETHNOBOTANICAL PROFILE OF TEHSIL KOTLI SATTIAN, RAWALPINDI, PAKISTAN

RAHMATULLAH QURESHI
AND
HUMAIRA SHAHEEN

NOVINKA

*New York*

For permission to use material from this book please contact us:
Telephone 631-231-7269; Fax 631-231-8175
Web Site: http://www.novapublishers.com

## NOTICE TO THE READER



### LIBRARY OF CONGRESS CATALOGING-IN-PUBLICATION DATA

The ethnobotanical profile of Tehsil Kotli Sattian, Rawalpindi, Pakistan / editors: Rahmatullah Qureshi and Humaira Shaheen.

p. cm.
Includes bibliographical references and index.
ISBN: 978-1-62417-290-8 (soft cover)
1. Ethnobotany--Pakistan--Rawalpindi District. I. Qureshi, Rahmatullah. II. Shaheen, Humaira.
GN476.73.E825 2013
581.6'309549142--dc23

2012048208

*Published by Nova Science Publishers, Inc. † New York*

# CONTENTS

# LIST OF TABLES

# LIST OF FIGURES

# List of Plates

# FOREWORD

In the course of the last couple of decades there has been an ever increasing interest and relevant research in the ethnobotany of Pakistan. In particular regions of northern and other parts of Pakistan that have been difficult to access are now, in part, being studied. Dr. Rahmatullah Qureshi has been one of the central figures in researching and documenting the ethnobotanical knowledge of areas of Pakistan that have not been studied earlier. This present book is a testimony of that. The ethnobotany of Kotli Sattian deals mainly with the vegetation and flora and its uses from the study area. Based on field work and knowledge from local people it documents the uses of plants for fodder, fuel, medicine, etc. A large proportion of people in this area, as with many rural areas in Pakistan, depend on plant life for all needs of livelihood, from building houses, fodder for livestock, to providing fuel for cooking and heating, and medicines. And as well as is found in many rural areas, the knowledge for use of plants for medicinal purposes is being lost without documentation. This is alarming, as age old recipes and identification of medicinal plants is being lost without any research into the efficacy and usefulness of the plants for possible future use. I hope this book will find a good use and add to our knowledge of not only the livelihoods of people in rural areas, but also help with the documentation of the floral wealth of Pakistan.

Shahina A. Ghazanfar
Head of Temperate Regional Team
Herbarium, Library, Art & Archives Directorate
Royal Botanic Gardens Kew,
Richmond, Surrey TW9 3AB. UK
September 20, 2012

# PREFACE

Kotli Sattian is an extension of the Indo-Himalayan ecological region that falls in the transitional zone of Irano-Saharan ecosystem which extends to the Southwest. This is a very beautiful subdivision of district Rawalpindi (Punjab), Pakistan which is bounded on the Northwest by the Murree Hills and touches the Kashmir territory on the East by bridging the river Jhelum on the Southwest. The evergreen hills with its gentle and steep slopes, the cool, fresh & health giving fountains, the dancing & singing streams, the winding roads & paths that wind through the lively pine woods and the bracing climate beckon the wanderer to this hilly area. The territorial name is derived from the mountain town of Kotli and the Satti tribe. This book is an amazing compilation of the plants used by the Satti tribe for medicine, food, forage, fuel, fiber and a multitude of other things. The authors have spent more than five years and documented more than 200 plant species valued by the inhabitants. This is undoubtedly the most substantial ethnobotanical survey ever undertaken, preserving indigenous knowledge of native flora for the future. This book is a valuable source for botanists, plant taxonomists, anthropologists, ethnobotanists, ethnopharmacologists, ecologists, foresters & range managers, nature lovers and tourists interested in the culture of Satti people and their way of use of native plants.

*Chapter 1*

# INTRODUCTION

## 1.1. THE STUDY AREA

Kotli Sattian is a subdivision of district Rawalpindi (Punjab) which is bounded on the east by the river Jhelum, on the south by Kahuta, southwest by the river Soan and on the North West by the Murree Hills (see Map). Kotli Sattian touches the Kashmir territory on the east and by bridging the river Jhelum on the south west. The evergreen hills with its gentle and steep slopes, the cool, fresh & health giving fountains, the dancing & singing streams, the winding roads & paths that wind through the lively pine woods and the bracing climate beckon the wanderer to this hilly area. The different areas of the sub division vary greatly from each other. The territorial name is derived from the mountain town of Kotli and the Satti tribe.

## 1.2. CLIMATE

Overall, the climate of study area is subtropical to temperate as well as sub-humid and falls in semi-arid region. The climate of the subdivision differs considerably on account of difference in altitude and height. It has four distinguishable seasons. Spring starts in March and ends in middle of May. In general, the climate is cold in winter and mild in summer; weheras, in valleys, it is mild. On the whole, the subdivision experiences cool and mild summers and cold winters. Summer spans from middle of May to June end of August. Winter starts in December and normally brings snowfall. The area receives 990 mm annual rainfall with an average of 254 mm monthly rainfall in the monsoon season (July-September). The temperature ranges were from 25°F to 117°F (Anonymous, 2008c). The mean relative humidity varies between 59-67% during the same period.

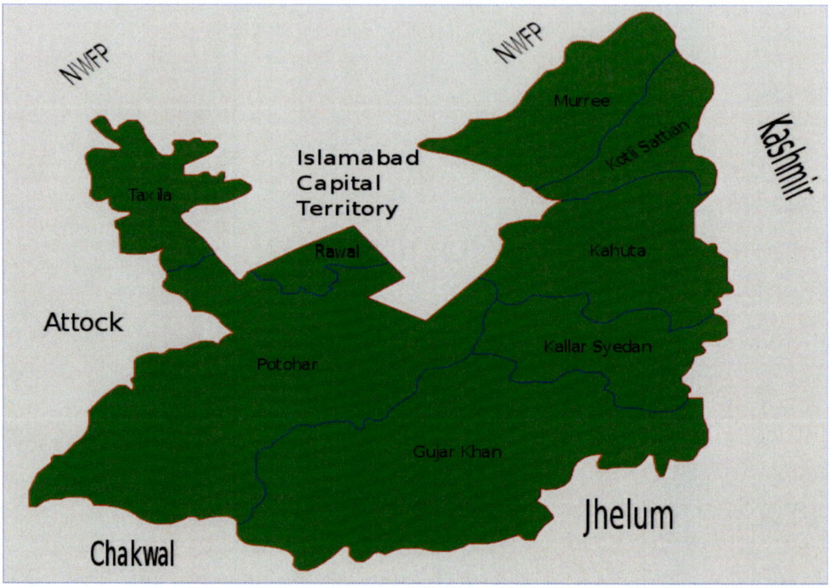


Location map of the study area.

### 1.2.1. Topography and Geology

Topography of the area is rocky and stoney nature with a range of elevation from 465 m to 1600 m typically embracing of steep and gullies. Like Margalla Hills, the area is made up of limestone and shale with a very thin layer of top soil. Like upland mountains, the study area is being affected by deforestation and during heavy rains; the top soil flows into the stream on the low lying areas and is slowly obstruction the natural surface drainage.

### 1.2.2. Soil

Since the soils are produced by the decaying action of the rocks through the weathering process, their texture and chemical contents vary from area to area. The soil of the hilly area including the sub mountain tract is residual and has been derived from parent materials of sandstone and purple red shale. The soil can be classified into three groups; soil on hilltops, soil on slopes and soil in terraces. On the hilltops, the soil occurs as a thin layer of silt clay or calcareous clay ranging in depth from two or three feet. They contain a high percentage of organic components due to the decay of material derived from thick coverage of vegetation. The soil deposits on the slopes are to some extent derived from the underlying rudimentary rocks. They are mainly colluvium in origin due to the steepness of the slopes. The soil contains a high percentage of decayed and un-decayed organic matter. The soil present on the terrace is mainly derived from the terrace material and the upper slopes of the hill. It is normally a mixture of boulders, pebbles, sand, silt and clay (Satti, 2000).

### 1.2.3. Scenery

Two main vegetation types are found in the study area. Low hilly ranges as well as foothills are dominated by scrub vegetation covered by *Dodonaea viscosa, Carissa opaca, Adhatoda zeylanica, Olea ferruginea* and *Punica granatum,* whereas the top hills covered by deciduous forest mixed with evergreen trees such as *Pinus roxburghii, Juglans regia, Grewia optiva, Quercus incana, Ficus carica, Pyrus pashia,* giving extremely beautiful and attractive look.

Plate 1. Scenic view of the study area.

## 1.3. FLORA AND VEGETATION

### 1.3.1. Flora

Like the Margalla Hills National Park, Kotli Sattian is an extension of Indo-Himalayan ecological region that falls in transitional zone of Irano-Saharan ecosystem much extending to south west. The buffer zonal area represents mixed vegetation type due to overlapping of two ecological regions. Thin film layered soils located in southern aspect of the hills coupled with sporadic rainfall force to develop scrub forest armed with xeric adaptation. On the other hand, the northern aspect located on more than 1600 meter above sea level (asl) having cool temperature and high precipitations dominated by evergreen forest mixed with deciduous forests having broad-leaved and evergreen trees. The most common trees, shrubs and herbs of the park are listed as follows:

**Table 1. Common floral elements found in the study area**

| Habit | Botanical Name | Family |
|---|---|---|
| Trees | *Acacia modesta* Wall. | Mimosaceae |
| | *A. nilotica* (L.) Delile. | Mimosaceae |
| | *Ailanthus altissima* (Mill.) Swingle | Simarubaceae |
| | *Bauhinia variegata* L. | Caesalpinaceae |
| | *Cassia fistula* L. | Caesalpinaceae |
| | *Emblica officinalis* Gaertn. | Euphorbiaceae |
| | *Olea ferruginea* Royle. | Oleaceae |
| | *Pinus roxburghii* Saragent. | Pinaceae |
| | *Pisticia integrimma* Stew ex Brandis | Anacardiaceae |
| | *Quercus leucotrichophora* A. Camus. | Fagaceae |
| Shrubs | *Adhatoda zeylanica* Nees | Acanthaceae |
| | *Carissa opaca* Stapf. ex Haines | Apocynaceae |
| | *Coleebrokia oppositifolia* Sm. | Lamiaceae |
| | *Dodonaea viscosa* (L.) Jacq. | Sapindaceae |
| | *Debregesia salicifolia* (D. Don) Rendle | Rhamnaceae |
| | *Grewia tenax* (Forssk.) Aschers & Schweinf. | Tiliaceae |
| | *Maytenus royleanus* (Wall. ex Lawson) Cufo. | Celastraceae |
| | *Otostagia limbata* (Bth.) Boiss. | Lamiaceae |

| Habit | *Botanical Name* | Family |
|---|---|---|
| | *Punica granatum* L. | Punicaceae |
| | *Rubus fruiticosus* Aag. | Rosaceae |
| | *Woodfordia fruticosa* (L.) S. Kurz *Olea ferrugina* L. | Lathyraceae |
| | *Zanthoxyllum armatum* DC. | Rutaceae |
| | *Zizyphus oxyphylla* Edgew. | Rhamnaceae |
| Herbs | *Alternanthera pungens* Kunth. | Amaranthaceae |
| | *Cannabis sativa* L. | Cannabinaceae |
| | *Colchicum aitchisonii* (Hook. f.) E. Nasir | Liliaceae |
| | *Dicliptera roxburghiana* Nees | Acanthaceae |
| | *Digera muricata* (L.) Mart. | Amaranthaceae |
| | *Geranium rotundifolium* L. | Geraniaceae |
| | *Malvastrum coromendelianum* L. | Malvaceae |
| | *Medicago polymorpha* L. | Fabaceae |
| | *Mentha longifolia* (L.) Huds | Lamiaceae |
| | *Micromeria biflora* (Ham.) Bth. | Lamiaceae |
| | *Plantago lanceolata* L. | Plantaginaceae |
| | *Rananculus sceleratus* L. | Rananculaceae |
| | *Salvia plebeia R. Br.* | Lamiaceae |
| | *Stellaria media* (L.) Cry. | Caryophyllaceae |
| | *Tagetes minuta* L. | Asteraceae |
| | *Taraxacum officinale* Weber. | Asteraceae |
| | *Trifolium repens* L. | Fabaceae |
| Climber | *Cardiospermum halicacabum* L. | Sapindaceae |
| | *Clematis montana* Buch. | Rananculaceae |
| | *Clematis napaulensis* Royle | Rananculaceae |
| | *Convolvulus arvensis* L. | Convulvulaceae |
| | *Gallium aparine* L. | Rubiaceae |
| | *Hedera nepallensis* K. Koch | Araliaceae |
| | *Lathyrus aphaca* L. | Fabaceae |
| | *Rhynchosia minima* (L.) DC | Fabaceae |
| | *Rubia cordifolia* L. | Rubiaceae |
| | *Smilax aspera* L. | Smilacaceae |
| | *Tinospora cordifolia* auct. non (DC). Miers | Menispermaceae |
| | *Vicia faba* L. | Fabaceae |

### 1.3.2. Vegetation Zones of Kotli Sattian

In general five distinct vegetation zones are identified in the project area as follows:

Plate 2. Scrub forest.

#### *1.3.2.1. Subtropical Scrub Forest*

This zone is found dominated by the plant community comprising on *Dodonaea viscosa*, *Adhatoda zeylanica* and *Carissa opaca*. The other associated shrubs include *Olea ferruginea, Acacia modesta, Punica granatum* and *Sageretia thea*. The main tree species of this zone are *Cassia fistula*, *Lannea coromandelica*, *Bauhinia variegata*, *Grewia optiva*, *Mallotus philipiensis,* etc. The common associated grasses are *Bothriochloa pertusa*, *Heteropon contortus*, *Pennisetum orientale* and *Themeda ananthera*.

#### *1.3.2.2. Transitional Zone*

This zone is situated above the elevation of scrub forest about 700 m comprising of large number of evergreen and deciduous plant species. The canopy is relatively thick. The understory and shrub cover is also fairly thick and is dominated by large number of unpalatable mostly evergreen plants with small leaves such as *Dodonaea viscosa, Carissa opaca, Adhatoda zeylanica*,

*Woodfordia fruticosa*, *Maytenus royleanus* and *Flacourtia indica*. The tree cover is dominated by *Mallotus philippensis*, whereas, *Cassia fistula* sub-dominated along with other associated species such as *Bauhinia variegata, Bombax ceiba, Albizia lebbeck* and *Lannea coromandelica*. Most of the trees are part of subtropical forest and pleasantly growing mostly on cooler aspect with sparse distribution. This reveals their preferences to more mesic conditions or shady slopes rather than steep hot slopes. *Pistacia chinensis* was found at high altitude, where more moisture was available.

### *1.3.2.3. Subtropical Pine Forest*

This zone is characterized by *Pinus roxburghii* as dominant species, whereas *Myrsine africana, Woodfordia fruticosa, Berberis lyceum* and *Carissa opaca* as co-dominant species in this forest. This habitat is located at the highest slopes mostly 900 m and above. This zone sustains large number of herbaceous plants species due to presence of plentiful moisture. The understory is sparse, especially the shrub layer and in some sites, under composed pine needles from a thick mat covering the forest floor. *Rumex hastatus* is one of the commonest shrubs growing on the rocks in this zone. Amongst the other characterstic species that are found in this habitat includes *Rubus ellipticus*. Typical herbaceous plants include V*iola canescens, Adiantum incisum, Galium elegans, Geranium rotundifolium* and *Duchesnea indica*. Among the grasses, *Themeda anathera* and *Heterropogon contortus* are well represented. This zone covers hilly, moderate to moderately steep slopes, ridge crests and steep slopes and is characterized by soils with low value of pH, with the altitude ranged between 930 m and 1230 m on north and north-western aspects.

### *1.3.2.4. Mixed Vegetation Zone*

This zone is situated on higher altitudes along the moderate and steep slopes at various aspects. Various species such as *Rosa brunonii, Punica granatum, Quercus leucotrichophora, Aesculus indica, Carissa opaca, Dodonaea viscosa, Woodfordia fruitcosa* and *Debregeasia salicifolia* are the characteristic features of this habitat. The area is dominated by *Pistacia chinensis, Zanthoxyllum armatum* and *Berberis lyceum* with a fairly continuous low scrub growth of xerophytic shrubs such as *Myrsine africanum* and the presence of mainly sub-tropical pine forest species. *Pinus roxburghii* was fairly distributed tree along with its associated trees such as *Pyrus pashia* and *Mallotus philippensis*. The herb layer is composed of variety of annuals and perennials that include *Adiantum incisum, Duchesnea indica, Viola canescens, Plantago lanceolata, Dicilptera bupleuroides* and *Micromeria biflora*.

Plate 3. Transitional zone.

Plate 4. Pine forest.

Plate 5. Mixed forest.

## 1.4. Socio-Economic Features

The dwellers are economically poor and dependent on livestock rearing and growing wheat on small scale especially on terraces. The forest area is under immense pressure due to overexploitation for fuel source (fuel wood extraction and trade) and trade of medicinal and economic plants. The livestock of the park area includes goats, cows, buffaloes and donkeys. The domesticated animals freely graze/browse and therefore causing overgrazing, which is another important factor responsible for limiting the growth of palatable grasses, forbs and herbs. People even collect and store fodder species from forests and cultivated areas for unavailabe period. Various fruits and ornamental species are grown by the inhabitants to meet their requirements. The list of such species is provided in Table 2.

**Table 2. Important tree species grown in Kotli Sattian**

| Fruit types | Botanical Name | Family |
|---|---|---|
| Almond | *Prunus amygdalus* Stockes | Rosaceae |
| Apple | *P. malus* L. | Rosaceae |
| Apricot | *P. armeniaca* L. | Rosaceae |
| Batangi | *P. pashia* | Rosaceae |
| Walnut | *Juglans regia* L. | Juglandaceae |
| Banana | *Musa sapientum* L. | Musaceae |
| Khatkal | *Citrus medica* L. | Rutaceae |
| Lemon | *C. limon* L. | Rutaceae |
| Loquat | *Eriobotrya japonica* Lindl. | Rosaceae |
| Mulberry (black) | *Morus nigra* L. | Moraceae |
| Mulberry (white) | *M. alba* L. | Moraceae |
| Peach | *Prunus persica* Seib and Zuce | Rosaceae |
| Sweet lime | *Citrus limetetioides* Tan. | Rutaceae |
| Sweet orange | *C. reticulate* Blanco | Rutaceae |
| Amlok | *Diospyros lotus* | |
| Japani phal | *D. kaki* | |

## 1.5. The National Park

The study area is included into Murree-Kotli Sattian-Kahuta National Park by the Government of Punjab vide their notification No. SOFT (EXT) XII-2/2009 dated 15-09-2009. The area inhabit various important wildlife species such as Kaleej pheasant, Common leopard, Grey goral, Barking deer, Grey and Black partridges, Rhesus Monkey and Asiatic Jackal (Anonymous, 2009).

Protection of biodiversity is the main purpose for the establishment of a national park however, being biologically, geologically and culturally important. The national parks also serve to cater for education, recreation and scientific purposes for the public. People visit parks to see and share the wonders of their land and to learn about the forces and the people who have shaped it through the centuries. Outstanding scenery, majestic places, natural flora and fauna in natural state renders positive impacts on visitors' mind thus helping divert a nation's attitude towards a healthy tract. National parks ensure the preservation of national natural heritage, culture and monuments, and present them to the public (Anonymous, 2009).

### 1.5.1. Role of National Parks in Nature Conservation

Due to the growing threats to nature such as global warming, population expansion, habitat destruction and worldwide reduction of biological diversity, national parks are now being considered as the ecological laboratories, gene pools and bulwarks against the irreversible change or the loss of species and hence, preserving the nature and ensuring the ecological health of the planet. National parks are the ideal places for eco-tourism, tracking, bird and animal watch and nature photography. National parks with their lush green forest components neutralize the carbon emissions in the surrounding environments.

## 1.6. Threats to Plants Diversity

### 1.6.1. Human Dwellings

Human settlement is the causative agent for the gradual declining of natural habitats. Furthermore, natural vegetation is used to fulfill the daily life requirements of the villagers. The local inhabitants extensively use plant resources for fuel wood, forage, medicinal and other customary use that resulted into decrease of economic plants population in the study area. The rural population is mostly illiterate and not aware about the significance of biodiversity and its sustainability.

### 1.6.2. Fire

Fires are very frequently observed in the project area that causes the damage of vegetation over large area. Like the Margalla Hills, the area faces about 85% fire during the May-June dry period preceding the monsoon rains. Most of the fire occurs on the upper slope on southern aspects.

Plate 6. Human settlement in the study area.

### 1.6.3. Grazing

The area faces heavy grazing that damage natural vegetation. When cattle walk over seedlings, the regeneration of forests becomes difficult. The domesticated animals like cattle, goats, buffaloes, sheep and donkeys compete with each other for food and space.

Plate 7. Grazing in the study area.

### 1.6.4. Invasive Plant Species

The introduction of exotic species like *Broussonetia papyrifera* and *Lantana camara* has distressed the natural vegetation, while *Parthenium hysterophorus* has accidentally disturbed different sites of the study area. The paper mulberry (*B. papyrifera*) is native to upper Burma and China and was introduced in Islamabad to make the capital green in 1960 (Stewart 1972). This species has very fast growth rate and its seed are dispersed through birds. It possesses allelopathic effects and drastically reduce ground flora except few species such as *Malvastrum coromendilianum, Cannabis sativa* and *Gallium aparine*. Moreover, this species is found responsible for pollen allergy.

## 1.7. ETHNOBOTANY

The human history on this planet will never be computed without a look at the rock of planets. The ancient record of many hundreds of plants species used by the human beings shows their importance in health, economy, shelter, clothing and food. Till last decade, peoples were more interested in surgery and allopathic medicines. They were dependant on plant resources for the treatment of different diseases. However, with the rise of resistant against antibiotics recently, there has been a reawakened interest in medicinal plants throughout the world (Anonymous, 2008a).

The traditional communities have indigenous knowledge of plants growing nearby their settlements. They have in borne knowledge of plants of the people of a particular culture and region. The aim of ethnobotany is to record the indigenous knowledge of plants used by the inhabitants of the area in question. It works with the collaboration between local communities and scientists to explore the potential use of native species. John Harshberger in 1896 first time defined the term ethnobotany to study the plants used by the primitive and aboriginal people. The definition of ethnobotany can be summed up in four words i.e. people, plant, interactions and uses (Hazart *et al.*, 2007).

Ethnobotany is the study of how the people of a particular culture and region make the use of indigenous plants. While the ethnobotanist explores how plants are used plants as food, shelter, medicine, clothing, hunting and in religious ceremonies. It is the science which studies 'the relationship between a given society and its environment and in particular the plant world'. Ethnobotany moved on the understanding of knowledge system through the use of anthropological methods and the work (Ram, 2004). Today, ethnobotany experienced a shift from the raw compilation of data to a greater methodological reorientation. This is also the beginning of academic ethnobotany. Ethnobotany requires a verity of skills: botanical training for the identification and preservation of plant specimens; anthropological training to understand the cultural concepts around the perception of plants; linguistic training, at least enough to transcribe local terms and understand native morphology, syntax, and semantics (Anonymous, 2008b).

Ethnobotanical knowledge is one of the parts of cultural heritage of an area that depicts the life-style and relationship of local community with their environmental and their mutual interaction which has profound importance for long term survival of people and biodiversity in an ecosystem. Generally, ethnobotanical studies involve an interaction between plants and people and foremost among these are the management of plant diversity by indigenous communities and the traditional use of medicinal plants. Medicinal plants provide health security to millions of rural people all over the world. According to WHO, over 80% of people in developing countries depend on traditional medicines for their primary health requirements (Farnsworth and Soejarto, 1991).

Principally, ethnobotanical works serve as preliminary method of research, suitable for gathering information on the use of plants. It has been proven again the 'quack' medical knowledge handed down by the common people constitutes sources of information useful for scientific research and that many plants utilized exclusively in popular tradition, when exposed under scientific examination, have been found to be useful for different sectors in the industry, therefore science and tradition have a strong connection between them, science in fact has often traditional origin (Lentini, 2000).

Today, ethnobotany is widely accepted as science of human interactions with plants and its ecosystem. Ethnobotany is multi-disciplinary science of botany, ecology and anthropology. Thus, ethnobotany is more than simply a study of plants useful to people. It is also devoted to understand the limitation and behavioral consequences of human population's action on their plant environment (Bhatti *et al.*, 2001).

Globally, this field is well recognized and a good amount of work has been done with reference to ethnobotany of various areas (Rout and Thatoi, 2009; Idu, 2009; Jain *et al.,* 2009; Albert, 2008; Mohan *et al.*, 2008; Srivastava, 2007; Anitha, 2008; Dobhal *et al.*, 2006; Ozbucak *et al.*, 2006; Ram, 2004; Ramana and Reddy, 2001; Oryem-Origa *et al.*, 1995. In Pakistan, this field is quite popular and a number of studies have been undertaken from different corners of the country (Qureshi and Bhatti, 2008; Ahmad *et al.*, 2008; Qureshi and Bhatti, 2006; Hussain *et al.*, 2008; Hazrat *et al.*, 2007; Hussain *et al.*, 2006; Islam *et al.*, 2006; Arshad *et al.*, 2005; Ashfaq *et al.*, 2004; Rafique *et al.,* 2004; Ahmed *et al.,* 2008; Arshad *et al.*, 2002; Qureshi *et al.*, 2001; Shinwari and Khan, 1999; Goodman and Ghafoor, 1992).

Young generations are paying less interest in experiences of old traditional knowledge. With the passage of time, this information is declining. Therefore, it is important to document this precious information about the uses of plants before its complete lost (Reddy *et al.*, 2005).

The area is a hilly tract and transitional zone in between subtropical to temperate zone resulting in unique floral biodiversity. Previously no study was carried out with respect to ethnobotany from the study area. Therefore, present study was designed to record the ethnobotanical uses of plants by the inhabitants of Kotli Sattian and adjacent its areas.

*Chapter 2*

# REVIEW OF LITERATURE

The use of plants is as old as human civilization. The term 'Ethnobotany' was coined by W. Harshberger in 1896 as uses of plants by primitive people. Today, this study encompasses all studies which concern the mutual relationship between plants and tradition people (Martin, 1995). Ethnobotanical field is well recognized throughout the world and a lot of work has been reported. Rasool (1998) reported 15 medicinally important plants from Saltanat of Oman. He described various parts of species which are used for the treatment of various diseases. Shaema *et al.* (2004) reported the traditional indigenous application of the plants of Parati valley of Kullu district, in western Himalaya. They recorded the uses of plants as fuel, fiber, timber, household articles and medicines by the local people of the study area. The ethnomedicinal properties of plants from Tian Mu Shan Biosphere Reserve of Zhejiang province, China has been reported by Chaudary *et al.* (2006). They discovered various folklore recipes of 67 medicinal plants.

The folk uses of medicinal plants of Calalabrin were recorded by Passaiacqua *et al.* in 2007. They identified 104 taxa distributed into 42 families. The ethnobotany of Nandi forest, Kenya has been reported by Jeruto *et al.* (2007). They identified 85 medicinal plants which are used for control of human ailments. Mohan *et al.* (2008) surveyed Tirunelveli district and he recorded a total of 80 species belonging to 72 genera and 46 families. Anitha *et al.*, (2008) reported the medico-ethnobotany of Chinnamayilaru, Periyamyilaru, region of Agasthiyamalai biosphere, Tamil Nadu, India. They reported wild plants of the project area which are used for skin treatment. Likewise, Ratnam and Raju (2008) studied the therapeutic properties of 21 crude drugs obtained from plants which are being used for bone fracture by the natives of Eastern Ghats. Vidyasagar and Prashantkumar (2007) reported the medicinal plants of Bidar district Karnataka, India and identified 18 plants species belonging to 13 families and 18 genera. Salazar *et al.* (2008) recorded herbal remedies used for gastrointestinal diseases from Maxico. They identified 85 different species which belong to 38 families. Likewise, Ugurlu and Secman (2008) discovered the medicinal plants of the Yunt Mountains. They identified 52 medicinal plants belonging to 32 families. Of them, 41 species were wild and 13 species were cultivated.

The ethnobotanical research is rich in India and a lot of work has been reported in this regard. Norscia and Tarli (2006) studied the ethnobotanical study of two forests area of Madagascar, India they reported 45 morphospiecies which is used as medicine. Pattanaik *et al.* (2008) reported the ethnobotanical plant mentioning 53 medicinal plants species belonging to 34 families and 52 different species in Malkangiri district Orissa, India. Koche *et al.*, (2008) reported the indigenous folk knowledge of the inhabitants of Nagzira, Maharashtra. They identified 70 different plant species having ethnobotanical uses. 32 families of angiosperm used by villager are taxonomically classified. Prasad *et al.* (2008) studies the ethnobotanical study of North Andaman, India. They identified 72 plant species and 67 genera and 43 families of Magnolioptyta from tropical rainforest.

The ethnobotany in Pakistan is going to be matured with the passage of time and various studies have been reported. Qureshi *et al.*, (2001) studied ethnomedicinal used on *Aloe barbadensis* in Nara desert. They have mentioned the traditional uses of this plant especially in digestive diseases and their livestock. Martin *et al.*, (2007) studies the Shogran valley (Hazara) and identified 77 species of herbs, 12 species of shrubs and 18 species of trees, which are used for medicinal purpose. Khan and Khatoon (2004) studied the ethnobotany of Haramosh and Bogrote valleys of Gilgit Agency, Northern Areas, and Pakistan. They identified 260 species and described 20 species of ethnobotanically importance. Qureshi *et al.*, (2006) studies the medicinal value of date palm in district Khairpur. They study the taxonomic description and record the indigenous knowledge for curing, processing and other native uses in various ways.

Ahmed *et al.* (2006) studied the traditional medicinal use of weed from district Attock. They identified 15 weeds species belonging to 15 genera of flowering plants. Hassan *et al.* (2007) reported the Ethnobotany and Phytochemistry of Lemon Grass. In the same way, Qureshi and Bhatti (2007) reviewed ethnobotany of wild gourd with from the Nara Desert, Pakistan.

They reported different recipes and taxonomic validation of this plant from the study area.

Ishtiaque (2006) described the indigenous plants in Samohni valley, A.J.K., Pakistan. He identified 36 species distributed in 26 families most of them are widely grown plants, are indigenous and herbs and there used for sexual disease and birth control.

The review clearly depicts that there is a lot of information exists on ethnobotanical discipline and there is broad spectrum scope of this study. Since 80% of the population of our country is living in villages, therefore there is great reliance on herbal medicines (Ikram and Hussain, 1978). Keeping in view, the present study was designed to record indigenous knowledge of plants used by the people of Santh Saroola, Kotli Sattian.

*Chapter 3*

# MATERIALS AND METHODS

## 3.1. COLLECTION OF ETHNOBOTANICAL INFORMATION

The indigenous knowledge of plants was extracted by using semi-structured questionnaire following the work of Qureshi (2004). Regular field trips were arranged in order to collect information about the folk knowledge of plants. During field trips, the questionnaire was used on random basis to interview the local inhabitants, including older people, herbalists (*Hakeems*), midwives (*Daai*) and elderly known people who were familiar with the traditional use of local plants. Analysis of data was carried out and indigenous knowledge was documented. In addition, literature was consulted (Gupta, 1945; Hill, 1952; Uphof, 1959; Niebuhr, 1970; Usher, 1974; Triska, 1975; Singh and Kachroo, 1976; Emboden, 1979; Weiner, 1980; Bean, 1981; Launert, 1981; Parmar and Kaushal, 1982; Lust, 1983; Baquar and Tasnif. 1984; Chiej, 1984; Grieve, 1984; Duke and Ayensu, 1985; Yeung, 1985; Chopra *et al.*, 1986; Anonymous, 1987; Polunin and Huxley 1987; Foster and Duke, 1990; Phillips and Rix, 1991; Coffey, 1993; Tsarong and Tsewang, 1994; Bown, 1995; Chevallier, 1996; Usmanghani *et al.*, 1997; Moerman, 1998; Manandhar, 2002; Shinwari *et al.*, 2007) to seek any noval use as well as authenticity of the use recorded from the study area.

## 3.2. SPECIMEN COLLECTION

Plant specimens were collected during the field trips from 2009 to 2012. The whole study area was thoroughly surveyed and a wide range of plant specimens were collected. All the specimens were dried and pressed using conventional method. These were mounted on herbarium sheets. The local people were interviewed to get vernacular names of plants. Flowering and fruiting period of species and ecology was recorded and such data was maintained in the field notebook.

## 3.3. Identification and Preservation of Specimens

The collected specimens were identified using floristic material (Nasir and Ali 1970-1989; Ali and Nasir 1989-1991; Ali and Qaiser, 1993-1995, 2000-2008; Jafri, 1966; Qureshi, 2004). These were compared with identified specimens in Herbarium of Pakistan (ISL), Quaid-e-Azam University Islamabad. The correctly identified specimens were deposited in the Department of Botany, Pir Mehr Ali Shah Arid Agriculture University for record.

## 3.4. Botanical Photographs

Botanical photographs are important supplement to the herbarium collection by providing information on habit of the recorded element. Photographs depicting habit of plants are provided for quick identification to botanists or other nature lovers.

## 3.5. Data Compilation

All species were arranged alphabetically and presented in the form Ethnobotanical Inventory. Using excel spreadsheet the data was statistically analyzed and certain graphical presentations were made.

*Chapter 4*

# RESULTS

This study documented 225 plant species belonging to 163 genera and 67 families which served to the natives for fulfilling daily life requirement. Of them, three ferns, one gymnosperm, 32 monocot and 189 dicot species are identified.

An aggregate list of floristic inventory along with local names, family and life form is provided in Table 3.

Most of the species were comprised of herbs (110 spp.) which contributed 48.46% of the ethnobotanical flora, followed by shrubs (44 spp., 19.38%), trees (26 spp., 11.45%) and grasses (22 spp., 9.69%), whereas remaining four habits were scarcely observed (Figure 1).

Seven major use categories were discovered from the study area and discussed as follows:

## 4.1. MEDICINAL PLANTS

The medicinal flora is habit/life form wise and family-wise alphabetically arranged in this inventory.

It includes: family, botanical name, vernacular name, part used, flowering period, medicinal properties and folk medicinal use.

## Table 3. Inventory of the flora along with their local names, family and habits

| S. No. | Plant Name | Local Name | Family Name | Habit/Life forms | | | | | | | | |
|---|---|---|---|---|---|---|---|---|---|---|---|---|
| | | | | Cl | Gr | Fe | He | Sh | Se | Sc | Pa | Tr |
| 1 | *Abutilon bidentatum* Hochst. ex Rich. | Kanghi buti | Malvaceae | – | – | – | – | √ | – | – | – | – |
| 2 | *Acacia modesta* Wall. | Phulai | Mimosaceae | – | – | – | – | – | – | – | – | √ |
| 3 | *Achyranthus aspera* L. | Puth kanda | Amaranthaceae | – | – | – | – | √ | – | – | – | – |
| 4 | *Adhatoda zeylanica* Medic. | Baikar | Acanthaceae | – | – | – | – | √ | – | – | – | – |
| 5 | *Adiantum capillus-veneris* L. | Persiaon shan | Adiantaceae | – | – | √ | – | – | – | – | – | – |
| 6 | *Adiantum incissum* | - | Adiantaceae | – | – | √ | – | – | – | – | – | – |
| 7 | *Ailanthus altissima* (Mill.) Swingle | Dhuravia | Simarubaceae | – | – | – | – | – | – | – | – | √ |
| 8 | *Ajuga bracteosa* Wall. ex Bth. | Kauri buti | Lamiaceae | – | – | – | √ | – | – | – | – | – |
| 9 | *Ajuga parviflora* Bth. | | Lamiaceae | – | – | – | √ | – | – | – | – | – |
| 10 | *Albizzia lebbeck* (L.) Benth. | Shrinh | Mimosaceae | – | – | – | – | – | – | – | – | √ |
| 11 | *Alternanthera pungens* Kunth. | Lundur | Amaranthaceae | – | – | – | √ | – | – | – | – | – |
| 12 | *Amaranthus hybridus* L. | Choleri | Amaranthaceae | – | – | – | √ | – | – | – | – | – |
| 13 | *Amaranthus spinosus* L. | Cholahi | Amaranthaceae | – | – | – | √ | – | – | – | – | – |
| 14 | *Amaranthus viridis* L. | Cholai | Amaranthaceae | – | – | – | √ | – | – | – | – | – |
| 15 | *Anagalis arvensis* L. | Bili booti | Primulaceae | – | – | – | √ | – | – | – | – | – |
| 16 | *Anisomeles indica* (L.) O. Kuntze | | Lamiaceae | – | – | – | √ | – | – | – | – | – |
| 17 | *Arabis himalaica* (Edgew.) O.E. Schulz | | Brassicaceae | – | – | – | √ | – | – | – | – | – |
| 18 | *Argyrolobium roseum* (Camb.) Jaub & Spach | | Fabaceae | – | – | – | √ | – | – | – | – | – |
| 19 | *Aristida cyanatha* Nees ex Steud. | | Poaceae | – | – | – | – | – | – | – | – | – |
| 20 | *Artemisia scoparia* Waldst. & Kit. | Dona Jhan | Asteraceae | – | – | – | √ | – | – | – | – | – |
| 21 | *Arundo donax* L. | Nara | Poaceae | – | √ | – | – | – | – | – | – | – |
| 22 | *Asparagus gracilis* Royle | Chou | Asparagaceae | – | – | – | – | – | – | √ | – | – |
| 23 | *Asphodelus tenuifolius* Cavan. | Bhagat/Piazi | Liliaceae | – | – | – | √ | – | – | – | – | – |
| 24 | *Astragalus squarrosus* Bunge | Kikri | Fabaceae | – | – | – | – | √ | – | – | – | – |
| 25 | *Avena fatua* L. | Jangli Jai | Poaceae | – | √ | – | – | – | – | – | – | – |
| 26 | *Bacopa monnieri* (L.) Wettst. | Jal nim | Scrophulariaceae | – | – | – | √ | – | – | – | – | – |

| S. No. | Plant Name | Local Name | Family Name | Habit/Life forms | | | | | | | | |
|---|---|---|---|---|---|---|---|---|---|---|---|---|
| | | | | Cl | Gr | Fe | He | Sh | Se | Sc | Pa | Tr |
| 27 | *Barleria acanthoides* Vahl | | Acanthaceae | – | – | – | √ | – | – | – | – | – |
| 28 | *Barleria cristata* L. | Bansa-Siya | Acanthaceae | – | – | – | √ | – | – | – | – | – |
| 29 | *Berberis lycium* Royle. | Sumbul | Berberidaceae | – | – | – | – | √ | – | – | – | – |
| 30 | *Boerhavia procumbens* Banks ex Roxb. | Itsit | Nyctaginaceae | – | – | – | √ | – | – | – | – | – |
| 31 | *Bombax ceiba* L. | Simbal | Bombacaceae | – | – | – | – | – | – | – | – | √ |
| 32 | *Calotropis procera* (Willd.) R.Br. | Aak, Mudar | Asclepiadaceae | – | – | – | – | √ | – | – | – | – |
| 33 | *Cannabis sativa* L. | Bhang | Cannabinaceae | – | – | – | √ | – | – | – | – | – |
| 34 | *Capsella bursa-pastoris* (L.) Medik | Shepherd's Purse | Brassicaceae | – | – | – | √ | – | – | – | – | – |
| 35 | *Cardiospermum halicacabum* L. | Kan phuti | Sapindaceae | √ | – | – | – | – | – | – | – | – |
| 36 | *Carissa opaca* Stapf ex. Haines | Garanda | Apocynaceae | – | – | – | – | √ | – | – | – | – |
| 37 | *Carthamus oxycantha* M. Bieb | Pohli | Asteraceae | – | – | – | √ | – | – | – | – | – |
| 38 | *Cenchrus ciliaris* L. | Dhaman | Poaceae | – | √ | – | – | – | – | – | – | – |
| 39 | *Cenchrus pennisetiformis* Hochst. & Steud. | Dhaman | Poaceae | – | √ | – | – | – | – | – | – | – |
| 40 | *Cenchrus setigerus* Vahl | Buth | Poaceae | – | √ | – | – | – | – | – | – | – |
| 41 | *Chenopodium album* L. | Bathu | Chenopodiaceae | – | – | – | √ | – | – | – | – | – |
| 42 | *Chenopodium ambrosioides* L. | Chandan Bathwa | Chenopodiaceae | – | – | – | √ | – | – | – | – | – |
| 43 | *Chrysopogon aucheri* (Boiss.) Stapf. | | Poaceae | – | √ | – | – | – | – | – | – | – |
| 44 | *Cirsium arvense* (L.) Scope. | Leh | Asteraceae | – | – | – | √ | – | – | – | – | – |
| 45 | *Clematis montana* Buch. | | Rananculaceae | √ | – | – | – | – | – | – | – | – |
| 46 | *Clematis napaulensis* Royle | | Rananculaceae | √ | – | – | – | – | – | – | – | – |
| 47 | *Colchicum aitchisonii* (Hook. f.) E. Nasir | Suranjan | Liliaceae | – | – | – | √ | – | – | – | – | – |
| 48 | *Colebrookia oppositifolia* Sm. | Shakar Danna | Lamiaceae | – | – | – | – | √ | – | – | – | – |
| 49 | *Commelina paludosa* Blume | Kanjura | Commelinaceae | – | – | – | √ | – | – | – | – | – |
| 50 | *Coniogramme rosthornii* Hieron. | | Coniogrammaceae | – | – | √ | – | – | – | – | – | – |
| 51 | *Convolvulus arvensis* L. | Hiran khuri, Leli | Convulvulaceae | √ | – | – | – | – | – | – | – | – |
| 52 | *Conyza aegyptica* Ait. | | Asteraceae | – | – | – | √ | – | – | – | – | – |
| 53 | *Conyza bonariensis* L. | Gidar buti | Asteraceae | – | – | – | √ | – | – | – | – | – |
| 54 | *Conyza canadensis* L. | Paleet | Asteraceae | – | – | – | – | √ | – | – | – | – |

**Table 3. (Continued)**

| S. No. | Plant Name | Local Name | Family Name | Habit/Life forms | | | | | | | | |
|---|---|---|---|---|---|---|---|---|---|---|---|---|
| | | | | Cl | Gr | Fe | He | Sh | Se | Sc | Pa | Tr |
| 55 | *Coronopus didymus* (L.) Sm. | Jangli haloon | Brassicaceae | – | – | – | √ | – | – | – | – | – |
| 56 | *Crotolaria medicagnea* Lam. | | Fabaceae | – | – | – | √ | – | – | – | – | – |
| 57 | *Cuscuta reflexa* Roxb. | Akash bail | Cuscutaceae | – | – | – | – | – | – | – | √ | – |
| 58 | *Cynodon dactylon* L. Pers | Khabbal ghaa | Poaceae | – | √ | – | – | – | – | – | – | – |
| 59 | *Cyperus rotundus* L. | Sedge Muther | Cyperaceae | – | – | – | – | – | √ | – | – | – |
| 60 | *Dactyloctenium aegyptium* L. | Gandeel | Poaceae | – | √ | – | – | – | – | – | – | – |
| 61 | *Datura stramonium* L. | Dhatura | Solanaceae | – | – | – | – | √ | – | – | – | – |
| 62 | *Desmostachya bipinnata* (L.) Stapf | Dab ghaa | Poaceae | – | √ | – | – | – | – | – | – | – |
| 63 | *Dicanthium annulatum* (Forssk.) Stapf | Murgha ghaa | Poaceae | – | √ | – | – | – | – | – | – | – |
| 64 | *Dicliptera roxburghiana* Nees | Somni | Acanthaceae | – | – | – | √ | – | – | – | – | – |
| 65 | *Digera muricata* (L.) Mart. | Tar rara | Amaranthaceae | – | – | – | √ | – | – | – | – | – |
| 66 | *Dioscorea deltoidea* Wall. ex Kunth | | Dioscoreaceae | √ | – | – | – | – | – | – | – | – |
| 67 | *Diospyros lotus* L. | | Ebenaceae | – | – | – | – | – | – | – | – | √ |
| 68 | *Dodonaea viscosa* (L.) Jacq. | Sanatha | Sapindaceae | – | – | – | – | √ | – | – | – | – |
| 69 | *Echinochloa crus-galli* (L.) P. Beauv. | Bara sawank | Poaceae | – | √ | – | – | – | – | – | – | – |
| 70 | *Echinops echinatus* Roxb. | Unt katara | Asteraceae | – | – | – | √ | – | – | – | – | – |
| 71 | *Eclipta prostrata* (L.) L. | Bhangra | Asteraceae | – | – | – | √ | – | – | – | – | – |
| 72 | *Equisetum ramosissimum* (Desf.) | | Equistaceae | – | – | √ | – | – | – | – | – | – |
| 73 | *Eragrostis ateroviens* (Desf.) Trin. ex Nees | | Poaceae | – | √ | – | – | – | – | – | – | – |
| 74 | *Eragrostis minor* Host. | Kusum | Poaceae | – | √ | – | – | – | – | – | – | – |
| 75 | *Eruca sativa* L. | Tara mira | Brassicaceae | – | – | – | √ | – | – | – | – | – |
| 76 | *Euphorbia clarkeana* Hkf. | | Euphorbiaceae | – | – | – | √ | – | – | – | – | – |
| 77 | *Euphorbia granulata* Forssk. | Sheer Bar | Euphorbiaceae | – | – | – | √ | – | – | – | – | – |
| 78 | *Euphorbia helioscopia* Mewski. | Chattri dodak | Euphorbiaceae | – | – | – | √ | – | – | – | – | – |
| 79 | *Euphorbia hirta* L. | Dudhi | Euphorbiaceae | – | – | – | √ | – | – | – | – | – |
| 80 | *Euphorbia indica* Lam. | | Euphorbiaceae | – | – | – | √ | – | – | – | – | – |

| S. No. | Plant Name | Local Name | Family Name | Habit/Life forms | | | | | | | | |
|---|---|---|---|---|---|---|---|---|---|---|---|---|
| | | | | Cl | Gr | Fe | He | Sh | Se | Sc | Pa | Tr |
| 81 | *Euphorbia prostrata* (L.) Ait | | Euphorbiaceae | – | – | – | √ | – | – | – | – | – |
| 82 | *Euphrasia himalayica* Wettst. | | Scrophulariaceae | – | – | – | √ | – | – | – | – | – |
| 83 | *Ficus bengalensis* L. | Barr | Moraceae | – | – | – | – | – | – | – | – | √ |
| 84 | *Ficus carica* L. | Anjeer | Moraceae | – | – | – | – | √ | – | – | – | – |
| 85 | *Ficus palmata* Forssk. | Phagwara | Moraceae | – | – | – | – | – | – | – | – | √ |
| 86 | *Ficus religiosa* L. | Peepal | Moraceae | – | – | – | – | – | – | – | – | √ |
| 87 | *Ficus roxburghii* Wall. ex Brand. | Dusi | Moraceae | – | – | – | – | – | – | – | – | √ |
| 88 | *Flacourtia indica* (Burm. f.) Merr. | Kakoh | Flacourtiaceae | – | – | – | – | – | – | – | – | √ |
| 89 | *Foeniculum vulgare* (Miller) | Sonf | Apiaceae | – | – | – | √ | – | – | – | – | – |
| 90 | *Fumaria indica* (Hausskn.) H.N. Pugsley | Shahtra, Papra | Fumariaceae | – | – | – | √ | – | – | – | – | – |
| 91 | *Gallium aparine* L. | | Rubiaceae | √ | – | – | – | – | – | – | – | – |
| 92 | *Geranium rotundifolium* L. | | Geraniaceae | – | – | – | √ | – | – | – | – | – |
| 93 | *Grewia optiva* Drum. ex. Burret. | Taman | Tiliaceae | – | – | – | – | – | – | – | – | √ |
| 94 | *Grewia tenax*(Forssk.) Aschers & Schweinf. | Gunger | Tiliaceae | – | – | – | – | √ | – | – | – | – |
| 95 | *Hedera nepallensis* K. Koch | | Araliaceae | √ | – | – | – | – | – | – | – | – |
| 96 | *Heliotopium europaeum* L. | Marand | Boraginaceae | – | – | – | √ | – | – | – | – | – |
| 97 | *Imperata cylindrica* (L.) Raeuschel | Cotton grass | Poaceae | – | √ | – | – | – | – | – | – | – |
| 98 | *Indigofera himalayensis* Ali | - | Fabaceae | – | – | – | – | √ | – | – | – | – |
| 99 | *Indigofera linifolia* (L. f) Retz | Torki | Fabaceae | – | – | – | √ | – | – | – | – | – |
| 100 | *Indigofera sessiliflora* DC | | Fabaceae | – | – | – | √ | – | – | – | – | – |
| 101 | *Ipomoea carnea* Jacq. | Akri | Convolvulaceae | – | – | – | – | √ | – | – | – | – |
| 102 | *Ipomoea hederacea* (L.) Jacq. | Wilayati Ak | Convolvulaceae | √ | – | – | – | – | – | – | – | – |
| 103 | *Ipomoea nil* (L.) Roth | | Convolvulaceae | √ | – | – | – | – | – | – | – | – |
| 104 | *Kickxia ramosissima* (Wall.) Janchen | | Scrophulariaceae | – | – | – | √ | – | – | – | – | – |
| 105 | *Lactuca auriculata* (Wall. ex Dc.) | | Asteraceae | – | – | – | √ | – | – | – | – | – |
| 106 | *Lactuca dissecta* D. Don. | | Asteraceae | – | – | – | √ | – | – | – | – | – |
| 107 | *Lactuca serriola* L. | | Asteraceae | – | – | – | √ | – | – | – | – | – |
| 108 | *Lannea coromandelica* L. | Kamlai | Anacardiaceae | – | – | – | – | – | – | – | – | √ |

## Table 3. (Continued)

| S. No. | Plant Name | Local Name | Family Name | Habit/Life forms | | | | | | | | |
|---|---|---|---|---|---|---|---|---|---|---|---|---|
| | | | | Cl | Gr | Fe | He | Sh | Se | Sc | Pa | Tr |
| 109 | *Lantana camara* L. | Panch phul | Verbenaceae | – | – | – | – | √ | – | – | – | – |
| 110 | *Lantana indica* Roxb. | | Verbenaceae | – | – | – | – | √ | – | – | – | – |
| 111 | *Lathyrus aphaca* L. | Jangali matar | Fabaceae | √ | – | – | – | – | – | – | – | – |
| 112 | *Launaea procumbens* (Roxb.) Ram. & Rajgo. | | Asteraceae | – | – | – | √ | – | – | – | – | – |
| 113 | *Lepidium sativum* L. | Haleon | Brassicaceae | – | – | – | √ | – | – | – | – | – |
| 114 | *Lotus corniculatus* (Wald. & Kit. ex Willd.) Briq. & Rech. F. | | Fabaceae | – | – | – | √ | – | – | – | – | – |
| 115 | *Mallotus philipensis* (Lam.) Muell. | Kameela | Euphorbiaceae | – | – | – | – | – | – | – | – | √ |
| 116 | *Malva neglecta* Waller. | Saunchal | Malvaceae | – | – | – | √ | – | – | – | – | – |
| 117 | *Malvastrum coromendelianum* L. | | Malvaceae | – | – | – | √ | – | – | – | – | – |
| 118 | *Maytenus royleanus* (Wall. ex Lawson) Cufodontis | Patakhi | Celastraceae | – | – | – | – | √ | – | – | – | – |
| 119 | *Medicago denticulata* Willd. | Maina | Fabaceae | – | – | – | √ | – | – | – | – | – |
| 120 | *Medicago laciniata* (L.) Mill. | | Fabaceae | – | – | – | √ | – | – | – | – | – |
| 121 | *Medicago polymorpha* L. | Maina | Fabaceae | – | – | – | √ | – | – | – | – | – |
| 122 | *Melia azedarach* L. | Bakain, Dharek | Meliacea | – | – | – | – | – | – | – | – | √ |
| 123 | *Melilotus indica* (L.) All. | Sinji | Fabaceae | – | – | – | √ | – | – | – | – | – |
| 124 | *Mentha longifolia* (L.) Huds | Pahari Poodina | Lamiaceae | – | – | – | √ | – | – | – | – | – |
| 125 | *Mentha royleana* Wall. ex Bth. | Jungli podina | Lamiaceae | – | – | – | √ | – | – | – | – | – |
| 126 | *Micromeria biflora* (Ham.) Bth. | | Lamiaceae | – | – | – | √ | – | – | – | – | – |
| 127 | *Mimosa rubicaulis* Lam. | Ral | Mimosaceae | – | – | – | – | √ | – | – | – | – |
| 128 | *Morus alba* L. | Tut | Moraceae | – | – | – | – | – | – | – | – | √ |
| 129 | *Morus nigra* L. | Tut | Moraceae | – | – | – | – | √ | – | – | – | – |
| 130 | *Myrsine africana* L. | Khokhal | Myrsinaceae | – | – | – | – | √ | – | – | – | – |
| 131 | *Nasturtim officinale* R. Br. | Pirya Halim | Brassicaceae | – | – | – | √ | – | – | – | – | – |
| 132 | *Nerium oleander* L. | Kanair | Apocynaceae | – | – | – | – | √ | – | – | – | – |
| 133 | *Nyctanthes arbor-tristis* L. | Harsingar | Oleaceae | – | – | – | – | √ | – | – | – | – |
| 134 | *Olea europaea* L. | | Oleaceae | – | – | – | – | – | – | – | – | √ |

| S. No. | Plant Name | Local Name | Family Name | Habit/Life forms | | | | | | | | |
|---|---|---|---|---|---|---|---|---|---|---|---|---|
| | | | | Cl | Gr | Fe | He | Sh | Se | Sc | Pa | Tr |
| 135 | *Olea ferruginea* Royle | Kahu | Oleaceae | – | – | – | – | √ | – | – | – | – |
| 136 | *Opuntia monacantha* Haw. | Naghphani | Cactaceae | – | – | – | – | √ | – | – | – | – |
| 137 | *Origanum vaulgare* L. | | Lamiaceae | – | – | – | √ | – | – | – | – | – |
| 138 | *Otostegia limbata* (Benth.) Boiss. | | Lamiaceae | – | – | – | – | √ | – | – | – | – |
| 139 | *Oxalis corniculata* L. | Khati buti | Oxalidaceae | – | – | – | √ | – | – | – | – | – |
| 140 | *Parthenium hytserophorus* L. | | Asteraceae | – | – | – | √ | – | – | – | – | – |
| 141 | *Peganum harmala* L. | Harmal | Zygophyllaceae | – | – | – | √ | – | – | – | – | – |
| 142 | *Periploca aphylla* Decne. | Batta | Asclepiadaceae | – | – | – | – | √ | – | – | – | – |
| 143 | *Phalaris minor* Retz. | Dhombi sitti | Poaceae | – | √ | – | – | – | – | – | – | – |
| 144 | *Phyla nodiflora* (L.) Greene | Bukan | Verbenaceae | – | – | – | √ | – | – | – | – | – |
| 145 | *Phyllanthus emblica* L. | Amla | Euphorbiaceae | – | – | – | – | – | – | – | – | √ |
| 146 | *Physalis divaricata* D.Don. | Aknaj | Solanaceae | – | – | – | √ | – | – | – | – | – |
| 147 | *Physalis minima* L. | | Solanaceae | – | – | – | √ | – | – | – | – | – |
| 148 | *Pinus roxburghii* Sargent | Chir | Pinaceae | – | – | – | – | – | – | – | – | √ |
| 149 | *Pistacia chinensis Bunge* | Kangar | Anacardiaceae | – | – | – | – | – | – | – | – | √ |
| 150 | *Plantago lanceolata* L. | Jangli isabgol | Plantaginaceae | – | – | – | √ | – | – | – | – | – |
| 151 | *Plantago major* L. | | Plantaginaceae | – | – | – | √ | – | – | – | – | – |
| 152 | *Plantago ovata* Frossk. | Isabgol | Plantaginaceae | – | – | – | √ | – | – | – | – | – |
| 153 | *Polygonum barbatum* L. | | Polygonaceae | – | – | – | – | √ | – | – | – | – |
| 154 | *Polygonum plebejum* R. Br. | | Polygonaceae | – | – | – | √ | – | – | – | – | – |
| 155 | *Polypogon fugax* Nees ex Steud. | | Poaceae | – | √ | – | – | – | – | – | – | – |
| 156 | *Pongamia pinnata*(L.) Pierre | Sukhchain | Fabaceae | – | – | – | – | – | – | – | – | √ |
| 157 | *Populus deltoides* Bartram ex Marsh. | Sufeida | Salicaceae | – | – | – | – | – | – | – | – | √ |
| 158 | *Potentilla supina* L. | Kanikar | Rosaceae | – | – | – | √ | – | – | – | – | – |
| 159 | *Psammogeton biternatum* Edgew. | --- | Apiaceae | – | – | – | √ | – | – | – | – | – |
| 160 | *Pteridium aquilinum* (L.) Kuhn | | Pteridaceae | – | – | √ | – | – | – | – | – | – |
| 161 | *Pueraria tuberosa* (Roxb. Ex Willd) DC. | Siali | Fabaceae | √ | – | – | – | – | – | – | – | – |
| 162 | *Punica granatum* L. | Daruna, Daruni | Punicaceae | – | – | – | – | √ | – | – | – | – |

**Table 3. (Continued)**

| S. No. | Plant Name | Local Name | Family Name | Habit/Life forms | | | | | | | | |
|---|---|---|---|---|---|---|---|---|---|---|---|---|
| | | | | Cl | Gr | Fe | He | Sh | Se | Sc | Pa | Tr |
| 163 | *Pyrus pashia* | Batangi | Rosaceae | – | – | – | – | – | – | – | – | √ |
| 164 | *Quercus dilatata* Lindley | Barungi | Fagaceae | – | – | – | – | – | – | – | – | √ |
| 165 | *Quercus incana* Roxb. | | Fagaceae | – | – | – | – | – | – | – | – | √ |
| 166 | *Quercus leucotrichophora* | Rein | Fagaceae | – | – | – | – | – | – | – | – | √ |
| 167 | *Rananculus muricatus* L. | Jal dhania | Rananculaceae | – | – | – | √ | – | – | – | – | – |
| 168 | *Rananculus sceleratus* L. | Jal dhania | Rananculaceae | – | – | – | √ | – | – | – | – | – |
| 169 | *Ranunculus arvensis* L. | Jal dhania | Ranunculaceae | – | – | – | √ | – | – | – | – | – |
| 170 | *Reinwardtia indica* Dumort. | Maraho | Linaceae | – | – | – | √ | – | – | – | – | – |
| 171 | *Rhynchosia minima* (L.) DC | | Fabaceae | √ | – | – | – | – | – | – | – | – |
| 172 | *Ricinis communis* L. | Arand | Euphorbiaceae | – | – | – | – | √ | – | – | – | – |
| 173 | *Rosa brunonii* Lindll. | Jangli gulab | Rosaceae | – | – | – | – | √ | – | – | – | – |
| 174 | *Rubia cordifolia* L. | Majeth | Rubiaceae | √ | – | – | – | – | – | – | – | – |
| 175 | *Rubus ellipticus* Smith | Aakhra | Rubiaceae | – | – | – | – | – | – | √ | – | – |
| 176 | *Rumex dentatus* L. | Jangli palak | Polygonaceae | – | – | – | √ | – | – | – | – | – |
| 177 | *Rumex hastatus* D. Don. | Khatimber/Chuki | Polygonaceae | – | – | – | – | √ | – | – | – | – |
| 178 | *Rumex nepalensis* Spreng | | Polygonaceae | – | – | – | √ | – | – | – | – | – |
| 179 | *Saccharum bengalense* Retz. | Kana | Poaceae | – | √ | – | – | – | – | – | – | – |
| 180 | *Saccharum spontaneum* L. | Kana | Poaceae | – | √ | – | – | – | – | – | – | – |
| 181 | *Sageretia thea* (Osback) M. C. Johnston | Gunger | Rhamnaceae | – | – | – | – | √ | – | – | – | – |
| 182 | *Salvia moorcroftiana* Wall. ex Bth. | Kali jarri. | Lamiaceae | – | – | – | √ | – | – | – | – | – |
| 183 | *Salvia plebeia R. Br.* | Sathi | Lamiaceae | – | – | – | √ | – | – | – | – | – |
| 184 | *Sauromatum venosum* (Ait.) Schott. | Samp di boti | Araceae | – | – | – | √ | – | – | – | – | – |
| 185 | *Saussurea albescens* (DC.) Schr. Bip. | | Asteraceae | – | – | – | √ | – | – | – | – | – |
| 186 | *Saussurea atkinsonii* (Clarke) | | Asteraceae | – | – | – | √ | – | – | – | – | – |
| 187 | *Saussurea heteromalla* DC. | Kali jiri | Asteraceae | – | – | – | √ | – | – | – | – | – |
| 188 | *Setaria glauca* (L) P.Beauv | Ban kangni | Poaceae | – | √ | – | – | – | – | – | – | – |

| S. No. | Plant Name | Local Name | Family Name | Habit/Life forms | | | | | | | | |
|---|---|---|---|---|---|---|---|---|---|---|---|---|
| | | | | Cl | Gr | Fe | He | Sh | Se | Sc | Pa | Tr |
| 189 | *Sida cordata* (Burm. f.) Borss.-Waalkes | | Malvaceae | – | – | – | √ | – | – | – | – | – |
| 190 | *Silene conoidae* L. | | Caryophyllaceae | – | – | – | √ | – | – | – | – | – |
| 191 | *Silybum marianum* (L.) Gaertin | Kandiari | Asteraceae | – | – | – | √ | – | – | – | – | – |
| 192 | *Sisymbrium irio* L. | Khub kalan | Brassicaceae | – | – | – | √ | – | – | – | – | – |
| 193 | *Smilax aspera* L. | | Smilacaceae | √ | – | – | – | – | – | – | – | – |
| 194 | *Solanum incanum* | Mahori | Solanaceae | – | – | – | – | √ | – | – | – | – |
| 195 | *Solanum nigrum* L. | Kach mach | Solanaceae | – | – | – | √ | – | – | – | – | – |
| 196 | *Solanum pseudocapsicum* L. | Jangli mirch | Solanaceae | – | – | – | – | √ | – | – | – | – |
| 197 | *Solanum surattense* Burm.f. | Kandiali | Solanaceae | – | – | – | – | √ | – | – | – | – |
| 198 | *Solanum villosum* (L.) Moench | Peelan | Solanaceae | – | – | – | √ | – | – | – | – | – |
| 199 | *Sonchus arvense* | | Asteraceae | – | – | – | √ | – | – | – | – | – |
| 200 | *Sonchus asper* (L.) Hill. | Dodak machal | Asteraceae | – | – | – | √ | – | – | – | – | – |
| 201 | *Sorghum bicolor* (L.) Moench. | Jawar/Chari | Poaceae | – | √ | – | – | – | – | – | – | – |
| 202 | *Sorghum halepense* (L.) Bern. | Baru | Poaceae | – | √ | – | – | – | – | – | – | – |
| 203 | *Stellaria media* (L.) Cry. | | Caryophyllaceae | – | – | – | √ | – | – | – | – | – |
| 204 | *Tagetes minuta* L. | Sadbarga | Asteraceae | – | – | – | √ | – | – | – | – | – |
| 205 | *Taraxacum officinale* Weber. | Dodak | Asteraceae | – | – | – | √ | – | – | – | – | – |
| 206 | *Taraxcum wallichii* DC. | | Asteraceae | – | – | – | √ | – | – | – | – | – |
| 207 | *Themeda anathera* (Nees) Hack | Loonder, Lunji | Poaceae | – | √ | – | – | – | – | – | – | – |
| 208 | *Trianthema portulacastrum* L. | Itsit | Aizoaceae | – | – | – | √ | – | – | – | – | – |
| 209 | *Tribulus terrestris* L. | Gokhru, Bhakhra | Zygophyllaceae | – | – | – | √ | – | – | – | – | – |
| 210 | *Trichodesma indicum* (L.) R. Br. | Gao zeban | Boraginaceae | – | – | – | √ | – | – | – | – | – |
| 211 | *Trifolium repens* L. | | Fabaceae | – | – | – | √ | – | – | – | – | – |
| 212 | *Urtica dioica* L. | Bichu boti | Urticaceae | – | – | – | √ | – | – | – | – | – |
| 213 | *Valeriana wallichii* DC. | | Valerianaceae | – | – | – | √ | – | – | – | – | – |
| 214 | *Verbascum thapsus* L. | Pahari tambako | Scrophulariaceae | – | – | – | √ | – | – | – | – | – |
| 215 | *Verbena officinalis* L. | | Verbenaceae | – | – | – | √ | – | – | | – | – |
| 216 | *Vicia faba* L. | Matri, Bakala | Fabaceae | √ | – | – | – | – | – | – | – | – |

**Table 3. (Continued)**

| S. No. | Plant Name | Local Name | Family Name | Habit/Life forms | | | | | | | | |
|---|---|---|---|---|---|---|---|---|---|---|---|---|
| | | | | Cl | Gr | Fe | He | Sh | Se | Sc | Pa | Tr |
| 217 | *Vicia sativa* L. | Rawari | Fabaceae | √ | – | – | – | – | – | – | – | – |
| 218 | *Viola cancescens* Wall. ex Roxb. | Banafsha | Violaceae | – | – | – | √ | – | – | – | – | – |
| 219 | *Vitex negundo* L. | Banna | Verbenaceae | – | – | – | – | √ | – | – | – | – |
| 220 | *Withania somnifera* (L.) Dunal. | Asghand/Aksan | Solanaceae | – | – | – | – | √ | – | – | – | – |
| 221 | *Woodfordia fruticosa* (L.) S. Kurz | Tavi | Vitaceae | – | – | – | – | √ | – | – | – | – |
| 222 | *Xanthium strumarium* L. | Bhurt | Asteraceae | – | – | – | – | √ | – | – | – | – |
| 223 | *Zanthoxylum alatum* Roxb. | Timbar | Rutaceae | – | – | – | – | √ | – | – | – | – |
| 224 | *Zizyphus mauritiana* Mill. | Beri, Jhari | Rhamnaceae | – | – | – | – | – | – | – | – | √ |
| 225 | *Zizyphus oxyphylla* Edgew. | Ber maloki | Rhamnaceae | – | – | – | – | √ | – | – | – | – |
| | | | | 16 | 22 | 5 | 109 | 44 | 1 | 2 | 1 | 26 |

Legend:

Cl= Climber, Gr= Grass, Fe= Fern, He= Herb, Sh= Shrub, Se= Sedge, Sc=Scandent, Pa= Parasite and Tr= Tree.

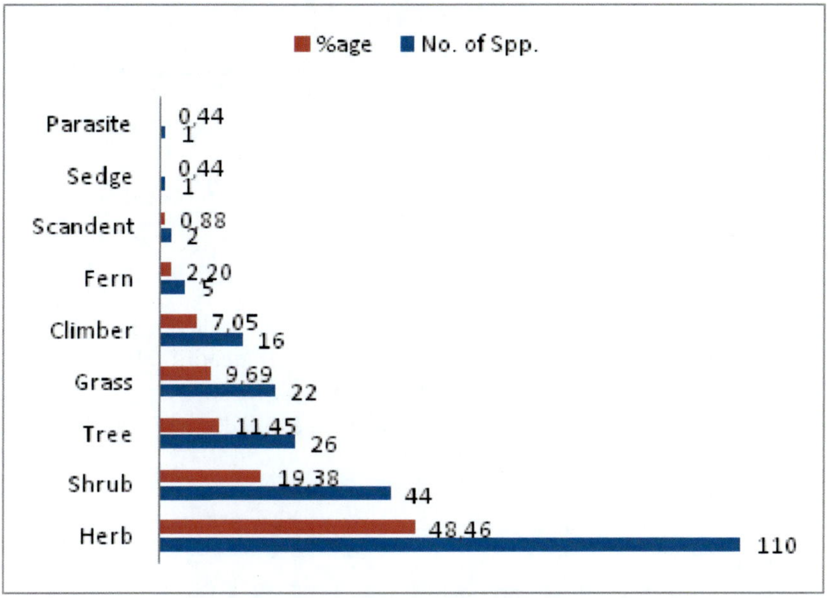


Figure 1. Habit of the ethnobotanical flora of the study area.

## 4.1.1. Climbers

| | | |
|---|---|---|
| | *Family: Araliaceae* | |
| 1 | *Botanical name:* | *Hedera nepalensis* K. Koch |
| | *Vernacular name:* | Bilri |
| | *Part used:* | Leaves |
| | *Flowering period:* | October - April |
| | *Medicinal properties:* | The leaves and the berries are cathartic, diaphoretic and stimulant. The decoction of the plant is given in skin diseases.( Manandhar, 2009) |
| | *Folk uses:* | ---- |
| | *Family: Convolvulaceae* | |
| 2 | *Botanical name:* | *Ipomoea carnea* Jacq. |
| | *Vernacular name:* | Wilayati Ak |
| | *Part used:* | Leaves |
| | *Flowering period:* | April - January. |
| | *Medicinal properties:* | ---- |

**(Continued)**

| | | |
|---|---|---|
| | *Folk medicinal use:* | Leaves are coated with sesame oil and applied as poultice for removing pain and inflammation. |
| | *Family: Fabaceae* | |
| 3 | *Botanical name:* | *Pueraria tuberosa* (Roxb. Ex Willd) DC. |
| | *Vernacular name:* | Siali |
| | *Part used:* | Roots |
| | *Flowering period:* | March-April |
| | *Medicinal properties:* | Root tuber is refrigerant, galactogogue, emetic, tonic and demulcent (Baquar 1989; Ali 1977; Nasir and Rafiq 1995; CSIR 2003). |
| | *Folk uses:* | The paste of tuber is applied to inflammation and joints pain. |
| 4 | *Botanical name:* | *Vicia sativa* L. |
| | *Vernacular name:* | Rawari |
| | *Part used:* | Whole plant, Seeds |
| | *Flowering period:* | March-May |
| | *Medicinal properties:* | The plant is used anti poisonous. Seeds are astringent and detergent and are used as a remedy for diarrhea (Ali 1977; Baquar 1989; Nasir and Rafiq 1995; Shinwari and Khan1998; CSIR 2003). |
| | *Folk uses:* | The powder of seeds is given to treat diarrhea. |
| | *Family: Rubiaceae* | |
| 5 | *Botanical name:* | *Rubia cordifolia* L. |
| | *Vernacular name:* | Majeth |
| | *Part used:* | Roots |
| | *Flowering period:* | March - May |
| | *Medicinal properties:* | Root is astringent, tonic and alterative. Stem is used in snake-bite scorpion string. The red dye Manjit is made from the root (Baqar, 1989) |
| | *Folk medicinal use:* | The powder of roots is given for regulating menstrual period. |
| | *Family: Sapindaceae* | |
| 6 | *Botanical name:* | *Cardiospermum halicacabum* L. |
| | *Vernacular name:* | Kan Phuti |
| | *Part used:* | Whole plant, leaves |
| | *Flowering period:* | January - October |

| | | |
|---|---|---|
| | *Medicinal properties:* | The whole plant is diaphoretic, diuretic, emetic, emmenagogue, laxative, refrigerant, rubefacient, stomachic and sudorific (Duke and Ayensu, 1985). It is for the treatment of rheumatism, nervous diseases, stiffness of the limbs and snakebite (Chopra *et al.*, 1986). The leaves are rubefacient and applied as a poultice in the treatment of rheumatism, earache (Chopra *et al.*, 1986; Baquar and Tasnif, 1984). |
| | *Folk medicinal use:* | Juice of leaves is slightly warmed and poured into ears to treat earache. |
| | *Family: Smilacaceae* | |
| 7 | *Botanical name:* | *Smilax aspera* L. |
| | *Vernacular name:* | ---- |
| | *Part used:* | Root |
| | *Flowering period:* | September-March |
| | *Medicinal properties:* | Root is diuretic (Baquar, 1989; CSIR, 2003; Ghazanfar, 1977). |
| | *Folk medicinal use:* | The paste of root is applied on wounds. |

## 4.1.2. Grasses

| | | |
|---|---|---|
| | *Family: Poaceae* | |
| 8 | *Botanical name:* | *Arundo donax* L. |
| | *Vernacular name:* | Narra, Kana |
| | *Part used:* | Whole plant |
| | *Flowering period:* | June - December. |
| | *Medicinal properties:* | The root infusion is regarded as antigalactagogue, depurative, diaphoretic, diuretic, emollient, hypertensive, hypotensive and sudorific (Duke and Wain, 1981). |
| | *Folk Medicinal use:* | Plant Boiled in wine with honey, the root or rhizome has been used for cancer. *Arundo* is reported to be used for indurations of the breast. |
| 9 | *Botanical name:* | *Cynodon dactylon* (L.) Pers |
| | *Vernacular name:* | Khabbal |

**(Continued)**

| | | |
|---|---|---|
| | *Part used:* | Whole plant |
| | *Flowering period:* | Throughout year |
| | *Medicinal properties:* | An infusion of the grass with milk is used for bleeding piles, irritation of urinary organs, dropsy and vomiting. The juice is also given in dysentery, ophthalmia, epilepsy and hematuria (Ali, 1977; Baquar, 1989; Nasir and Rafiq, 1995; Shinwari and Khan, 1998; CSIR, 2003). |
| | *Folk medicinal use:* | The paste of whole plant is applied on wounds to heal. |
| | *Family: Adiantatceae* | |
| 10 | *Botanical name:* | *Adiantum capillus-veneris* L. |
| | *Vernacular name:* | Persiaon shan |
| | *Part used:* | Whole plant |
| | *Flowering period:* | Throughout the year |
| | *Medicinal properties:* | The plant is used as emollient, demulcent and diuretic (Shinwari *et al.,* 2007). The fresh or dried leafy fronds are used as antidandruff, antitussive, astringent, demulcent, depurative, emetic, weakly emmenagogue, emollient, weakly expectorant, febrifuge, galactogogue, laxative, pectoral, refrigerant, stimulant, sudorific and tonic (Uphof, 1959; Usher, 1974; Launert, 1981; Lust, 1983; Chiej, 1984; Grieve, 1984; Duke and Ayensu, 1985; Chopra *et al.*, 1986; Foster & Duke, 1990). A tea or syrup is used in the treatment of coughs, throat afflictions and bronchitis (Foster and Duke, 1990). |
| | *Folk medicinal use:* | The plant mixed with *Glycyrrhiza glabra* and *Viola odorata* along with brown sugar (*Gur*) is boiled in water and given to treat flue, cough and asthma. |
| 11 | *Botanical name:* | *Adiantum incissum* Forsisa. |
| | *Vernacular name:* | ---- |
| | *Part used:* | Leaves |
| | *Flowering period:* | Throughout the year |

| | |
|---|---|
| *Medicinal properties:* | The leaves are used as a cure for cough, fever and chest affection. They are also employed for diabetes and externally the leaf paste is used to cure skin diseases. It has antibacterial properties also. Ethanolic (50%) extract of the plant is hypoglycemic and spasmogenic (Asolkar *et al.*, 1992). |
| *Folk medicinal use:* | The paste of leaves is applied externally for skin allergies. |

## 4.1.3. Herbs

| | | |
|---|---|---|
| | *Family: Amaranthaceae* | |
| 12 | *Botanical name:* | *Amaranthus hybridus* L. |
| | *Vernacular name:* | Choleri |
| | *Part used:* | Leaves |
| | *Flowering period:* | February - June |
| | *Medicinal properties:* | A tea made from the leaves is astringent. It is used in the treatment of intestinal bleeding, diarrhea and excessive menstruation (Huxley, 1992). |
| | *Folk medicinal use:* | Leaves are used as potherb for treating constipation. |
| 13 | *Botanical name:* | *Amaranthus viridis* L. |
| | *Vernacular name:* | Cholai |
| | *Part used:* | Whole plant, leaves |
| | *Flowering period:* | February - June |
| | *Medicinal properties:* | Leaves emollient, used in scorpion sting and snakebite (Baquar and Tasnif, 1984; Shinwari *et al.*, 2007). |
| | *Folk medicinal use:* | The plant is cooked as vegetable and used against constipation. |
| | *Family: Apiaceae* | |
| 14 | *Botanical name:* | *Foeniculum vulgare* Miller |
| | *Vernacular name:* | Soonuf |
| | *Part used:* | Whole plant, seeds |
| | *Flowering period:* | June - August |

**(Continued)**

| | |
|---|---|
| *Medicinal properties:* | Fennel has herbal use for remedy digestive problems.The seeds, leaves and roots can be used, but the seeds are most active medicinally and are the part normally used. An essential oil is often extracted from the fully ripened and dried seed for medicinal use, though it should not be given to pregnant women. An infusion is used in the treatment of indigestion, abdominal distension, stomach pains etc. It helps in the treatment of kidney stones and, when combined with a urinary disinfectant like Arctostaphylos uva-ursi, makes an effective treatment for cystitis. It can also be used as a gargle for sore throats and as eyewash for sore eyes and conjunctivitis.Fennel is often added to purgatives in order to allay their tendency to cause gripe, and also to improve the flavour. An infusion of the seeds is a safe and effective cure for wind in babies. An infusion of the root is used to treat urinary disorders. An essential oil obtained from the seed is used in aromatherapy. The essential oil is bactericidal, carminative and stimulant. Some caution is advised, see notes above on toxicity. ( Phillips & Foy, 2009), (Grieve, 2009), (Genders, 2009) |
| *Folk medicinal use:* | The fruit is given in eye sight and stomach pain. Seeds are roasted in fire and used in gastric problems and obesity. Plant extract is used as cooling agent in the form syrup (*Sharbat*). The seed of plant mixing with equal quantity of the seeds of *Carum carvi* (*Kala zeera*), table salt and lemon juice is boiled in water that is given in vomiting, indigestion and pain in abdomen. The seeds along with *Carum carvi* seed, black salt and molasses (*gur*) are boiled in water to obtain decoction which is given to the cattle after giving birth of baby for the healing of wounds. Seeds are chewed for digestive and inflammation as cooling agent for mouth wash and throat pain. Seeds grinded and made into pills that are used as digestive and appetizer. |

| | | |
|---|---|---|
| | *Family: Asteraceae* | |
| 15 | *Botanical name:* | *Launaea procumbens* (Roxb.) Ram. & Rajgo. |
| | *Vernacular name:* | Dodak |
| | *Part used:* | Whole plant |
| | *Flowering period:* | October - December |
| | *Medicinal properties:* | Not reported |
| | *Folk medicinal use:* | The paste of eaves mixed with the leaves of *Lawsonia inermis* is applied on skin rashes. |
| 16 | *Botanical name:* | *Silybum marianum* (L.) Gaertin (30) |
| | *Vernacular name:* | Kandiari |
| | *Part used:* | Leaves, whole plant, seeds |
| | *Flowering period:* | February - May |
| | *Medicinal properties:* | The whole plant is astringent, bitter, cholagogue, diaphoretic, diuretic, emetic, emmenagogue, hepatic, stimulant, stomachic and tonic (Grieve, 1984; Lust, 1983; Anonymous, 1987; Bown, 1995). It is used internally in the treatment of liver and gall bladder diseases, jaundice, cirrhosis, hepatitis and poisoning (Bown, 1995). Seeds are used in fever, cancer, uterine troubles, diabetes, antispasmodic and jaundice (Shinwari *et al.*, 2007). |
| | *Folk Medicinal use:* | The powder of seeds is given in jaundice. |
| 17 | *Botanical name:* | *Tagetes minuta* L. |
| | *Vernacular name:* | Sadbarga |
| | *Part used:* | Whole plant |
| | *Flowering period:* | August - November |
| | *Medicinal properties:* | The whole plant is anthelmintic, antispasmodic, aromatic, diaphoretic, diuretic, purgative and stomachic (Usher, 1974; Bown, 1995). It is used internally in the treatment of gastritis, indigestion and internal worms. Externally, it is used to treat haemorrhoids and skin infections (Bown, 1995). |
| | *Folk Medicinal use:* | Decoction of flowers is used to cure chronic fever. The flowers are burnt in oil and that is poured in ear or juice of leaves is poured in ear to cure earache. |

**(Continued)**

| | | |
|---|---|---|
| | *Family:Boraginaceae* | |
| 18 | *Botanical name:* | *Heliotropium europaeum* L. |
| | *Vernacular name:* | Hathi sundi |
| | *Part used:* | leaves |
| | *Flowering period:* | January-April. |
| | *Medicinal properties:* | ---- |
| | *Folk medicinal use:* | Leaves are burnt in sesame oil and applied on skin irritation. |
| 19 | *Botanical name:* | *Trichodesma indicum* (L.) R. Br. |
| | *Vernacular name:* | Gao Zeban |
| | *Part used:* | Leaves |
| | *Flowering period:* | August - October |
| | *Medicinal properties:* | ---- |
| | *Folk Medicinal use:* | Leaves or even whole plant is made into powder and added with sugar which is given to cure bloody motions, abdominal pain and as cooling agent. |
| | *Family: Brassicaceae* | |
| 20 | *Botanical name:* | *Eruca sativa* L. |
| | *Vernacular name:* | Tara Meera |
| | *Part used:* | Whole plant, oil |
| | *Flowering period:* | April-June. |
| | *Medicinal properties:* | ---- |
| | *Folk Medicinal use:* | Oil obtained from seeds is applied externally for removing pimples, leprosy, skin dandruff, etc. the oil is boiled and after cooling given to livestock to expel abdominal worms. This oil is also used for cooking vegetables. Plant is cooked in gurh and made into balls. These balls are given to cow to thicken the milk. |
| 21 | *Botanical name:* | *Sisymbrium irio* L. |
| | *Vernacular name:* | Khub Kalan |
| | *Part used:* | Whole plant, seeds |
| | *Flowering period:* | March-May. |
| | *Medicinal properties:* | The seeds are expectorant, restorative and stimulant. They are used in the treatment of asthma. Externally, they are used as a stimulating poultice. Thereby wash foreign matter out of the eye (Moerman, 1998). |

| | | |
|---|---|---|
| | | An infusion of the leaves is used in treating affections of the throat and chest The seeds have been placed under the lids of sore eyes in order to cause weeping (Chopra *et al.*, 1986). Expectorant, asthma and throat and chest infection (Shinwari *et al.,* 2007). |
| | *Folk Medicinal use:* | Seeds are boiled in water and juice is used for treating chronic fever. |
| | *Family: Chenopodiaceae* | |
| 22 | *Botanical name:* | *Chenopodium album* L. |
| | *Vernacular name:* | Bathu |
| | *Part used:* | Whole plant |
| | *Flowering period:* | January - September |
| | *Medicinal properties:* | The plant is laxative and anthelmintic used in hepatic disorders and enlarged spleen. The root is given in jaundice, urinary diseases and rheumatic. Fruits and roots are supposed to have antidote to snakebite (Baquar and Tasnif, 1984). |
| | *Folk Medicinal use:* | The plant is used as potherb and given to constipating patients to relieve. |
| 23 | *Botanical name:* | *Chenopodiumam ambrosioides* L. |
| | *Vernacular name:* | Chandan bathu |
| | *Part used:* | Leaves, whole plant |
| | *Flowering period:* | April - January. |
| | *Medicinal properties:* | ---- |
| | *Medicinal properties:* | The plant is used as an anthelmintic. Oil is effective against round and hookworms and also for piles (Shinwari *et al.*, 2007). |
| | *Folk medicinal use:* | Leaves are crushed and applied on headache. |
| | *Family: Lamiaceae* | |
| 24 | *Botanical name:* | *Ajuga bracteosa* Wall. ex Bth. |
| | *Vernacular name:* | Kauri buti |
| | *Part used:* | Leaves |
| | *Flowering period:* | March - December. |
| | *Medicinal properties:* | The plant is bitter, astringent and tonic. Leaves are used in fever as a substitute of *Cinchona* (Shinwari *et al.*, 2007). |
| | *Folk medicinal use:* | The use of plant as vegetable seems to be useful in constipation and piles. It is supposed as cooling effect. |

**(Continued)**

| | | |
|---|---|---|
| 25 | *Botanical name:* | *Mentha longifolia* (L.) Huds. |
| | *Vernacular name:* | Sufaid Poodina |
| | *Part used:* | Whole plant, leaves |
| | *Flowering period:* | May - November. |
| | *Medicinal properties:* | Juice of leaves anthelmintic, diuretic and emmenagogue (Baquar and Tasnif, 1984). The leaves and flowering stems are antiasthmatic, antispasmodic, carminative and stimulant (Niebuhr, 1970; Grieve, 1984; Chopra et al., 1986; Usmanghani et al., 1997). A tea made from the leaves has traditionally been used in the treatment of fevers, headaches, digestive disorders and various minor ailments (Foster and Duke, 1990). |
| | *Folk medicinal use:* | The decoction of plant is given in stomach problems like nausea, vomiting, diarrhea and dysentery. The powder of the plant is also given for the same purpose. The decoction of dried leaves, sonf (*Foeniculum vulgare*) and salt is given to treat vomiting. |
| 26 | *Botanical name:* | *Mentha royleana* Wall. ex Bth. |
| | *Vernacular name:* | Jungli podina |
| | *Part used:* | Aerial parts, leaves |
| | *Flowering period:* | June - August |
| | *Medicinal properties:* | The dried leaves are made into powder and used with curd as stomachic agent and carminative in diarrhea and dysentery (Shinwari and Khan 1998; CSIR 2003; Hedge 1990). |
| | *Folk medicinal use:* | The powder of leaves mixed in curd is given to treat diarrhea. The fresh leaves are crushed with chilies and garlic and mixed with curd which is used to increase appetite. |
| 27 | *Botanical name:* | *Origanum vaulgare* L. |
| | *Vernacular name:* | ----- |
| | *Part used:* | Whole plant |
| | *Flowering period:* | Throughout the year |
| | *Medicinal properties:* | Oregano has long been used as a medicinal herb in treatment of various conditions: fevers, diarrhea, indigestion, jaundice and vomiting. |

| | | |
|---|---|---|
| | | Recent studies have shown that Oregano displays anti-oxidant, anti-fungal and antibiotic properties. Due to its anti-oxidant functions, Oregano could become helpful agent in treatment of cancer, heart disease and high blood pressure. It is also useful as a digestive aid, since it promotes salivation. Used externally, Oregano is successful in treatments of rheumatism, muscle and joint pain, sores and swellings. Oregano oil can help combat toothache (Baquar, 1989) |
| | *Folk medicinal use:* | Leaves are used for making tea used as an appetizer. |
| | *Family: Lamiaceae* | |
| 28 | *Botanical name:* | *Salvia moorcroftiana* Wall. ex Bth. |
| | *Vernacular name:* | Kali jarri |
| | *Part used:* | Root, seed and leaves |
| | *Flowering period:* | May-June |
| | *Medicinal properties:* | Leaves are used as a remedy for guinea worms, itches and are applied to wounds as poultice. Seeds are emetic used in hemorrhoids, dysentery and colic and is applied to boils. Root is used in cough and cold (Baquar 1998; Hedge 1990; Nasir and Rafiq 1995; CSIR 2003). |
| | *Folk Medicinal use:* | The paste of leaves slightly warmed mixing with Brassica oil and applied on inflamed parts. |
| 29 | *Botanical name:* | *Salvia plebeia* R. Br. |
| | *Vernacular name:* | Sathi |
| | *Part used:* | Seeds and leaves |
| | *Flowering period:* | July-August |
| | *Medicinal properties:* | The herb is employed as diuretic, astringent and anthelmintic. Seeds are stimulant, rubefacient, carminative, expectorant, demulcent, refrigerant, and diuretic, abotifacient, aphrodisiac, antipoison, mucilaginous and nutritive; used in diarrhea, piles, gonorrhea, leucorrhoea, seminal weakness, menorrhea and hemorrhoids. Leaves are used in toothache (Baquar 1998; Hedge 1990; Nadkarni 1970; Kirtikar and Basu 1933; CSIR 2003). |

**(Continued)**

| | | |
|---|---|---|
| | *Folk Medicinal use:* | The powder of seeds is given in diarrhea. |
| | *Family: Liliaceae* | |
| 30 | *Botanical name:* | *Colchicum aitchisonii* (Hook. f.) E. Nasir |
| | *Vernacular name:* | Suranjan shireen |
| | *Part used:* | Bulb and whole plant |
| | *Flowering period:* | February-April |
| | *Medicinal properties:* | Plant is used for the treatment of gout and rheumatism. It is also has been studied as an anticancer agent (saroya, 2011). |
| | *Folk medicinal use:* | Powder of bulb is dusted on wounds to heal. |
| | *Family: Linaceae* | |
| 31 | *Botanical name:* | *Reinwardtia indica* Dumort. |
| | *Vernacular name:* | Marhoo |
| | *Part used:* | Whole plant and leaves |
| | *Flowering period:* | March-April |
| | *Medicinal properties:* | The plant is used in the treatment of paralysis. Crushed leaves and stems are applied to wounds infested with maggots in cattle (Abdulla 1972a; Baquar 1989; Nasir and Rafiq 1995; CSIR 2003). |
| | *Folk medicinal use:* | The paste of leaves is applied on wounds. |
| | *Family: Malvaceae* | |
| 32 | *Botanical name:* | *Malva neglecta* Waller. (102) |
| | *Vernacular name:* | Sonchal |
| | *Part used:* | Whole plant |
| | *Flowering period:* | Throughout the year |
| | *Medicinal properties:* | The plant is found useful for treating skin conditions, coughs and throat irritants (Anonymous, 2012). |
| | *Folk medicinal uses:* | Past of leaves are useful for skin problems. |
| 33 | *Botanical name:* | *Sida cordata* (Burm. f.) Borss.-Waalkes |
| | *Vernacular name:* | Bariar |
| | *Part used:* | Whole plant |
| | *Flowering period:* | Throughout the year |

| | | |
|---|---|---|
| | *Medicinal properties:* | Root is used as astringent, stomachic, diuretic, febrifuge and demulcent given in diarrhea. The juice of plant is used in rheumatism, gonorrhea and spermatorrhoea. Seeds are considered as aphrodisiac, laxative and demulcent recommended in gonorrhea, cystitis, colic and piles (Shinwari *et al.*, 2007). |
| | *Folk medicinal use:* | Juice of the plant is given in joints pain. |
| | *Family: Nyctaginaceae* | |
| 34 | *Botanical name:* | *Boerhavia procumbens* Banks ex Roxb. |
| | *Vernacular name:* | Itsit |
| | *Part used:* | Whole plant |
| | *Flowering period:* | August - September |
| | *Medicinal properties:* | Plant is expectorant, laxative, stomachic, diaphoretic, emetic, anthelmintic and febrifuge used in anemia, jaundice and edema. Juice extract is diuretic (Shinwari *et al.,* 2007). |
| | *Folk Medicinal use:* | The plant is cooked as vegetable as a diuretic agent to remove internal swellings. |
| | *Family: Oxalidaceae* | |
| 35 | *Botanical name:* | *Oxalis corniculata* L. |
| | *Vernacular name:* | Khati buti |
| | *Part used:* | Whole plant, leaves |
| | *Flowering period:* | March - December |
| | *Medicinal properties:* | The whole plant is anthelmintic, antiphlogistic, astringent, depurative, diuretic, emmenagogue, febrifuge, lithontripic, stomachic and styptic (Chopra *et al.*, 1986; Duke and Ayensu, 1986; Moerman, 1998). It is used in the treatment of influenza, fever, urinary tract infections, enteritis, diarrhea, traumatic injuries, sprains and poisonous snake bites. The juice of the plant, mixed with butter, is applied to muscular swellings, boils and pimples (Manandhar, 2002). The plant is a good source of vitamin C and is used as an antiscorbutic in the treatment of scurvy (Baquar and Tasnif, 1984; Chopra *et al.*, 1986). The leaf juice is applied to insect bites, burns and skin eruptions (Srivastava, 1989). |

**(Continued)**

| | | |
|---|---|---|
| | *Folk Medicinal use:* | Juice of plant is given dysentery. Leaves extract is taken by the patients in jaundice. |
| | *Family: Plantaginaceae* | |
| 36 | *Botanical name:* | *Plantago lanceolata* L. |
| | *Vernacular name:* | Ispaghol |
| | *Part used:* | Seeds |
| | *Flowering period:* | April - September |
| | *Medicinal properties:* | They have a bitter flavor and are astringent, demulcent, mildly expectorant, haemostatic and ophthalmic (Triska, 1975; Singh and Kachroo, 1976; Launert, 1981, Lust, 1983; Foster & Duke, 1990; Chevallier, 1996). Internally given in diarrhea, gastritis, peptic ulcers, irritable bowel syndrome, hemorrhage, hemorrhoids, cystitis, bronchitis, catarrh, sinusitis, asthma and hay fever (Baquar and Tasnif, 1984; Bown, 1995; Chevallier, 1996). They are used externally in treating skin inflammations, malignant ulcers, cuts, stings etc. (Grieve, 1984). The heated leaves are used as a wet dressing for wounds, swellings etc. (Weiner, 1980; Foster and Duke, 1990). The root is a remedy for the bite of rattlesnakes; it is used in equal portions with *Marrubium vulgare* (Coffey, 1993). The seeds are used in the treatment of parasitic worms (Weiner, 1980). Distilled water made from the plant makes an excellent eye lotion (Chiej, 1984). An extract of them has antibacterial properties (Chopra *et al.*, 1986). |
| | *Folk Medicinal use:* | Plant husk is used against in constipation and dysentery. Husk soaked in water is given in obesity. |
| 37 | *Botanical name:* | *Plantago major* L. |
| | *Vernacular name:* | Ispaghol |
| | *Part used:* | Seeds |
| | *Flowering period:* | August - September |

| | | |
|---|---|---|
| | *Medicinal properties:* | They have a bitter flavor and are astringent, demulcent, mildly expectorant, haemostatic and ophthalmic (Triska, 1975; Singh and Kachroo, 1976; Launert, 1981, Lust, 1983; Foster & Duke, 1990; Chevallier, 1996). Internally given in diarrhea, gastritis, peptic ulcers, irritable bowel syndrome, hemorrhage, hemorrhoids, cystitis, bronchitis, catarrh, sinusitis, asthma and hay fever (Baquar and Tasnif, 1984; Bown, 1995; Chevallier, 1996). They are used externally in treating skin inflammations, malignant ulcers, cuts, stings etc. (Grieve, 1984). The heated leaves are used as a wet dressing for wounds, swellings etc. (Weiner, 1980; Foster and Duke, 1990). The root is a remedy for the bite of rattlesnakes; it is used in equal portions with *Marrubium vulgare* (Coffey, 1993). The seeds are used in the treatment of parasitic worms (Weiner, 1980). Distilled water made from the plant makes an excellent eye lotion (Chiej, 1984). An extract of them has antibacterial properties (Chopra *et al.*, 1986). |
| | *Folk medicinal use:* | Plant husk is used against in constipation and dysentery. Husk soaked in water is given in obesity. |
| 38 | *Botanical name:* | *Plantago ovate* Frossk. |
| | *Vernacular name:* | Ispaghol |
| | *Part used:* | seeds |
| | *Flowering period:* | August - September |
| | *Medicinal properties:* | The dried seeds and the seed husks are demulcent, emollient and laxative (Uphof, 1959; Launert, 1981; Baquar and Tasnif, 1984; Grieve, 1984; Chevallier, 1996). They are used in the treatment of dysentery, catarrhal conditions of the genito-urinary tract, inflamed membranes of the intestinal canal etc (Hill, 1952; Uphof, 1959; Grieve, 1984) Husk is used as a demulcent and as a bulk laxative in the treatment of constipation, dysentery and other intestinal complaints, having a soothing and regulatory effect upon the system (Grieve, 1984). |
| | *Folk Medicinal use:* | Plant husk is used against in constipation and dysentery. Husk soaked in water is given in obesity. |
| | *Family: Polygonaceae* | |

**(Continued)**

| | | |
|---|---|---|
| 39 | *Botanical name:* | *Polygonum plebejum* R. Br. |
| | *Vernacular name:* | ----- |
| | *Part used:* | Whole plant |
| | *Flowering period:* | September - December |
| | *Medicinal properties:* | The dried and powdered plant is taken internally in pneumonia. The root is used in bowel complaints. The whole plant is used as an excellent remedy for cholera. (Shinwari and Khan, 1999). |
| | *Folk Medicinal use:* | Plant is used for making syrups (*Sharbat*) and prescribed to relieve fever and throat pain. This type of use is supposed having cooling effect. |
| 40 | *Botanical name:* | *Rumex dentatus* L. |
| | *Vernacular name:* | Jangli palak |
| | *Part used:* | Roots, leaves |
| | *Flowering period:* | March - May |
| | *Medicinal properties:* | The root is used antiseptic, cutaneous disorder and astringent (Chopra *et al.*, 1986). |
| | *Folk medicinal use:* | Leaves are used as potherb and given in constipation and piles. |
| 41 | *Botanical name:* | *Rumex nepalensis* Spreng. |
| | *Vernacular name:* | ----- |
| | *Part used:* | Leaves and root |
| | *Flowering period:* | March - May |
| | *Medicinal properties:* | The root is purgative (Chopra *et al.*, 1986, Manandhar, 2002). It is used as a substitute for rhubarb (Rheum spp.) (Chopra *et al.*, 1986). A strong decoction of the root is applied to dislocated bones. A paste of the root is applied to swollen gums (Manandhar, 2002). The leaves are used in the treatment of colic (Chopra *et al.*, 1986). The juice of the leaves is applied externally to relieve headaches. A decoction of the plant is used to wash the body in order to alleviate body pain (Manandhar, 2002). |
| | *Folk medicinal use:* | Leaves are antidote of *Urticaria dioica* (*Bichhoo buti*) and rubbed on affected parts caused by touching of the plants. |
| | *Family: Rananculaceae* | |
| 42 | *Botanical name:* | *Ranunculus muricatus* L. |

| | | |
|---|---|---|
| | *Vernacular name:* | Jal dhania |
| | *Part used:* | Leaves |
| | *Flowering period:* | |
| | *Medicinal properties:* | The plant is used is used in the treatment of intermittent fevers, gout and asthma (Manandhar, 2002). |
| | *Folk Medicinal use:* | The paste of leaves is applied on ringworm and athlete foot. |
| 43 | *Botanical name:* | *Ranunculus arvensis* L. (155) |
| | *Vernacular name:* | ----- |
| | *Part used:* | Leaves |
| | *Flowering period:* | March - April |
| | *Medicinal properties:* | ----- |
| | *Folk medicinal use:* | The paste of leaves is applied on ringworm and athlete foot. |
| | *Family: Rosaceae* | |
| 44 | *Botanical name:* | *Potentilla supina* L. |
| | *Vernacular name:* | Kanikar |
| | *Part used:* | Root |
| | *Flowering period:* | February - April |
| | *Medicinal properties* | Root is tonic, febrifuge and astringent (Baquar 1989; CSIR 2003). |
| | *Folk medicinal use:* | The powder of root is given with milk as tonic. |
| | *Family: Scrophulariaceae* | |
| 45 | *Botanical name:* | *Bacopa monnieri* (L.) Wettst. |
| | *Vernacular name:* | Jalnim, ambassador |
| | *Part used:* | Whole plant |
| | *Flowering period:* | January – May |
| | *Medicinal properties:* | Plant used as cardiac and nerve tonic, useful in insanity and epilepsy, asthma and hoarsness. Leaves and stalk diuretic and aperient (Baquar and Tasnif, 1984). |
| | *Folk medicinal use:* | The plant along with *Piper nigrum* and salt is crushed in water and the obtained juice is given to treat skin irritation and urticaria. |
| 46 | *Botanical name:* | *Verbascum thapsus* L. |
| | *Vernacular name:* | Pahari Tambaku |

**(Continued)**

| | | |
|---|---|---|
| | *Part used:* | Leaves, root and flowers |
| | *Flowering period:* | March - October |
| | *Medicinal properties:* | The leaves and the flowers are anodyne, anti-inflammatory, antiseptic, antispasmodic, astringent, demulcent, diuretic, emollient, expectorant and vulnerary (Uphof, 1959; Triska, 1975; Lust, 1983; Chiej, 1984; Grieve, 1984; Foster and Duke, 1990). An infusion is taken internally in the treatment of a wide range of chest complaints and also to treat diarrhoea (Grieve, 1984; Bown, 1995). Poultice of the leaves is a good healer of wounds and is also applied to ulcers, tumours and piles (Grieve, 1984; Foster and Duke, 1990; Chevallier, 1996). A decoction of the roots is said to alleviate toothache and also relieve cramps and convulsions (Grieve, 1984). Febrifuge, antilice, bactericide, wound, aphrodisiac and narcotic (Shinwari *et al.,* 2007). |
| | *Folk medicinal use:* | Roots are boiled in water and given in migraine. The decoction of leaves is used to remove body pain and also in sleeplessness. |
| | *Family: Solanaceae* | |
| 47 | *Botanical name:* | *Physalis divaricata* D. Don. |
| | *Vernacular name:* | Aknaj |
| | *Part used:* | Fruit and leaves |
| | *Flowering period:* | August – November |
| | *Medicinal properties:* | Fruit is diuretic, tonic, aperient and gonorrhea. Juice of leaves mixed with water and mustard oil is used as a cure for earache. Fruits are said to from an ingredient in medicinal oil given for spleen disorders (Baquar 1989; Nair 1985; CSIR 2003, Nasir and Rafiq, 1995). |
| | *Folk medicinal use:* | The fruit is chewed raw as mild laxative. |
| 48 | *Botanical name:* | *Solanum villosum* (L.) Moench |
| | *Vernacular name:* | Peelan/Kachmach |
| | *Part used:* | Whole plant |
| | *Flowering period:* | December-February |

| | | |
|---|---|---|
| | *Medicinal properties:* | Diuretic, expectorant, rheumatic, gonorrhea, heart diseases and skin diseases (Shinwari *et al.,* 2007). |
| | *Folk Medicinal use:* | Like *Solanum nigrum.* |
| | *Family: Valerianaceae* | |
| 49 | *Botanical name:* | *Valeriana wallichii* DC. |
| | *Vernacular name:* | Mushk bala |
| | *Part used:* | Root |
| | *Flowering period:* | March-May |
| | *Medicinal properties:* | ---- |
| | *Folk medicinal use:* | The powder of roots is given in diarrhea. |
| | *Family: Verbenaceae* | |
| 50 | *Botanical name:* | *Verbena officinalis* L. |
| | *Vernacular name:* | ----- |
| | *Part used:* | Fresh leaves |
| | *Flowering period:* | June - Dec. |
| | *Medicinal properties:* | Not reported. |
| | *Folk Medicinal use:* | Fresh leaves are crushed and applied on rheumatic and joints pain. |
| 51 | *Botanical name:* | *Phyla nodiflora* (L.) Greene |
| | *Vernacular name:* | Bukan buti |
| | *Part used:* | Leaves and young shoots. |
| | *Flowering period:* | Throughout the year |
| | *Medicinal properties:* | Leaves and young shoots are sometimes used in curing indigestion in children. Its decoction is considered as cooling agent and demulcent in cases of venereal diseases (Baquar 1989; Nasir and Rafiq 1995; Jafri and Ghafoor, 1974). |
| | *Folk medicinal use:* | The powder of aerial parts is used in bleeding piles. |
| | *Family: Violaceae* | |
| 52 | *Botanical name:* | *Viola cancescens* Wall. Ex Roxb. |
| | *Vernacular name:* | Banafsha |
| | *Part used:* | Whole plant, leaves, flowers |
| | *Flowering period:* | February - May |

**(Continued)**

| | | |
|---|---|---|
| | *Medicinal properties:* | Astringent, purgative, demulcent, constipation, nervous disorders and acne (Shinwari *et al.*, 2007). |
| | *Folk medicinal use:* | The decoction of leaves and flowers is given in throat pain, flue, cough and asthma. |
| | *Family: Zygophyllaceae* | |
| 53 | *Botanical name:* | *Peganum harmala* L. |
| | *Vernacular name:* | Harmal |
| | *Part used:* | Leaves, fruit, seeds and whole plant |
| | *Flowering period:* | April - October. |
| | *Medicinal properties:* | The fruit and seed are digestive, diuretic, hallucinogenic, narcotic and uterine stimulant (Emboden, 1979; Bown, 1995). The same are given internally in treating stomach complaints, urinary and sexual disorders, epilepsy, menstrual problems, mental and nervous illnesses (Bown, 1995). The seed has also been used as an anthelmintic in order to rid the body of tapeworms (Chopra *et al.*, 1986). The seeds contain the substance 'harmine' which is being used in research into mental disease, encephalitis and inflammation of the brain. The oil obtained from the seed is said to be aphrodisiac, galactogogue, ophthalmic, soporific and vermifuge. The oil is also said to have properties (Emboden, 1979). The seed is used externally in the treatment of hemorrhoids and baldness (Bown, 1995). The whole plant is said to be abortifacient, aphrodisiac, emmenagogue and galactogogue. A decoction of the leaves is used in the treatment of rheumatism. The root has been used as a parasiticide in order to kill body lice (Chopra *et al.*, 1986). It is also used internally in the treatment of rheumatism and nervous conditions (Chevallier, 1996). |

| | | |
|---|---|---|
| | *Folk medicinal use:* | Aerial parts are grinded that is applied as poultice on pimples, boils and carbuncle. The plant along with salt and sugar (*Gur*) is burnt into fire and resultant fumes are given to the patience of small pox (*Khasra*). The same is also applied in the case of fever (*Parchhawan*). |
| 54 | *Botanical name:* | *Tribulus terrestris* L. |
| | *Vernacular name:* | Bhakhra |
| | *Part used:* | Whole plant, fruits and seeds |
| | *Flowering period:* | Almost throughout the year. |
| | *Medicinal properties:* | The seed is abortifacient, alterative, anthelmintic, aphrodisiac, astringent, carminative, demulcent, diuretic, emmenagogue, galactogogue, pectoral and tonic (Grieve, 1984; Duke and Ayensu, 1985; Yeung, 1985). A decoction is used in treating impotency in males, nocturnal emissions, gonorrhea and incontinence of urine (Grieve, 1984; Chopra *et al.*, 1986). It has also proved effective in treating painful urination, gout and kidney diseases (Chopra *et al.*, 1986). The plant has shown anticancer activity. The flowers are used in the treatment of leprosy. The stems are used in the treatment of scabious skin diseases and psoriasis. The dried and concocted fruits are used in the treatment of congestion, gas, headache, liver, ophthalmia and stomatitis (Duke and Ayensu, 1985). |
| | *Folk medicinal use:* | The powder of plant is given for regulating menstruation period. The whole plant is boiled in water and is given in constipation during pregnancy. It is also given to ease delivery. Its spiny fruit are dried and grounded in to powder then this powder is given 1/4$^{th}$ teaspoon with water in early morning by patients of jaundice early in the morning in the one fourth of teaspoon along with water. Fruit powder is taken orally to cure leucorrhoea and back ache. |

## 4.1.5. Parasite

| | *Family: Cuscutaceae* | |
|---|---|---|
| 55 | *Botanical name:* | *Cuscuta reflexa* Roxb. |
| | *Vernacular name:* | Akash Bail |
| | *Part used:* | Seeds and Whole plant |
| | *Flowering period:* | August - September |
| | *Medicinal properties:* | Anthelmintic, carminative, alterative, purgative and diuretic used in jaundice, joint pains, paralysis and vomiting (Baquar and Tasnif, 1984). |
| | *Folk medicinal use:* | Seeds are boiled in water and recommended in chronic fever. Plant is crushed by pistil and mortar. Three to four teaspoons of water is added and is used with or without salt by the patients of hepatitis. |

## 4.1.6. Scandent

| | *Family: Rosaceae* | |
|---|---|---|
| 56 | *Botanical name:* | *Rubus ellipticus* Smith |
| | *Vernacular name:* | Aakhra |
| | *Part used:* | Fruit |
| | *Flowering period:* | March - May |
| | *Medicinal properties:* | ---- |
| | *Folk medicinal use:* | The ripened fruit is use as a source of energy and tonic. |

## 4.1.7. Sedge

| | *Family: Cyperaceae* | |
|---|---|---|
| 57 | *Botanical name:* | *Cyperus rotundus* L. |
| | *Vernacular name:* | Dela |
| | *Part used:* | Root and leaves |
| | *Flowering period:* | June- September |
| | *Medicinal properties:* | ---- |
| | *Folk medicinal use:* | The rhizome and leaves *of Mentha arvesis* is boiled in water and given in cholera. |

### 4.1.8. Shrubs

| | *Family: Acanthaceae* | |
|---|---|---|
| 58 | *Botanical name:* | *Adhatoda zeylanica* Medic. |
| | *Vernacular name:* | Baikar |
| | *Part used:* | Leaves, flowers and Whole plant |
| | *Flowering period:* | January - April |
| | *Medicinal properties:* | The plant is used as emollient, demulcent and diuretic (Shinwari *et al.,* 2007). The fresh or dried leafy fronds are used as antidandruff, antitussive, astringent, demulcent, depurative, emetic, weakly emmenagogue, emollient, weakly expectorant, febrifuge, galactogogue, laxative, pectoral, refrigerant, stimulant, sudorific and tonic (Uphof, 1959; Usher, 1974; Launert, 1981; Lust, 1983; Duke and Ayensu, 1985; Chiej, 1984; Grieve, 1984; Chopra *et al.*, 1986; Foster & Duke, 1990). A tea or syrup is used in the treatment of coughs, throat afflictions and bronchitis (Foster and Duke, 1990). |
| | *Folk medicinal use:* | The powder of plant is administered in dysentery. The powder of leaves and flowers is given to diabetic. Leaves are crushed and boiled for 15/20 minutes to make the decoction which is used in jaundice, boils and burning feet acting as cooling agent.<br><br>The same is given to children before start of rainy season as preventive measure for pimples and boils. Leaves are chewed to cure throat pain and inflammation and in phthisis. Dried leaves are powdered and this powder is orally taken for blood purification. To cure pneumonia of cattle, the plant is given as fodder. Equal amounts of leaves of this plant, Bhang (*Cannabis sativa)* and Piyaz (*Allium cepa)* are crushed and given to cattle as appetizer. |

**(Continued)**

| | | |
|---|---|---|
| 59 | *Botanical name:* | *Achyranthes aspera* L. |
| | *Vernacular name:* | Puth kanda |
| | *Part used:* | roots, whole plant |
| | *Flowering period:* | September-December |
| | *Medicinal properties:* | The plant is purgative, diuretic, astringent and emetic used in dropsy, piles, skin eruptions and colic (Baquar and Tasnif, 1984). |
| | *Folk medicinal use:* | Branches/twigs are used as tooth stick (*Dantan*). Roots are made into pieces and tied in necklace form (*Mala and Tabiz*) to treat fever and to hasten the process of delivery. The inflorescence is burnt into fire and the obtained ash is mixed with sugar which is given orally to cure cough and asthma. |
| | *Family: Apocynaceae* | |
| 60 | *Botanical name:* | *Nerium oleander* L. |
| | *Vernacular name:* | Kanair |
| | *Part used:* | Leave, stem and root |
| | *Flowering period:* | April - October |
| | *Medicinal properties:* | Leaves flower and stem possess cardio-tonic properties. Bark is used in skin diseases especially leprosy. Root is used as abortifacient; paste is useful in scorpion sting and snake bite; also used in aphrodisiac formulations. Leaves are used in cutaneous eruptions; decoction of leaves is applied externally to reduce swellings and also to maggots infested wounds. Twigs are used as toothbrush against tooth decay with care (Baquar 1989; Nasir and Rafiq 1995; CSIR 2003; Shinwari and Khan, 1998). |
| | *Folk medicinal use:* | Root paste is applied on scorpion sting and snake bite. Twigs are used as toothbrush for relieving toothache and decay. |
| | *Family: Apocynaceae* | |
| 61 | *Botanical name:* | *Carissa opaca* Stapf ex. Haines |
| | *Vernacular name:* | Granda |
| | *Part used:* | Leaves, branches, root and fruit |
| | *Flowering period:* | April- June |

| | | |
|---|---|---|
| | *Medicinal properties:* | Fruit is cardiac stimulant and leaf decoction is given in asthma (Shinwari *et al.*, 2007). |
| | *Folk medicinal use:* | Decoction of leaves is used to cure hepatitis, flue and headache. Root is boiled in water or dipped in water over night, and this decoction is taken to cure diabetes. Leaves and spines are crushed and made powder given for curing jaundice. |
| | *Family: Asclepiadaceae* | |
| 62 | *Botanical name:* | *Calotropis procera* (Willd.) R. Br. |
| | *Vernacular name:* | Aak, Madar |
| | *Part used:* | Leaves and root |
| | *Flowering period:* | Throughout the year |
| | *Medicinal properties:* | Root and bark are used as tonic, sudorific, alterative, antispasmodic, stomachic, digestive, expectorant and emetic (Qureshi, 2004). |
| | *Folk Medicinal uses:* | The leaves are coated with mustard oil and slightly warmed on fire and tied on boils. Latex of leaves is externally applied on boils and wounds. Root is ground with equal proportions of sugar (*Gurh*) and made into small pills, these are taken with water to cure rheumatic pain. The ash of root is mixed with honey and used in cough. |
| | *Family: Asclepiadaceae* | |
| 63 | *Botanical name:* | *Periploca aphylla* Decne. |
| | *Vernacular name:* | Baata |
| | *Part used:* | Bark and latex |
| | *Flowering period:* | March – May |
| | *Medicinal properties:* | Plant is said to be useful in cerebral fever and is stomachic. Bark decoction is purgative. Milky juice is locally applied to tumors and swellings (Baquar 1989; Ali, 1983; CSIR, 2003). |
| | *Folk medicinal use:* | Milky latex is applied on boils. |
| | *Family: Berberidaceae* | |
| 64 | *Botanical name:* | *Berberis lyceum* Royle. |
| | *Vernacular name:* | Sumblu |
| | *Part used:* | Root, leaves, whole plant |
| | *Flowering period:* | April - June. |

**(Continued)**

| | | |
|---|---|---|
| | *Medicinal properties:* | The roots are aperient, carminative, febrifuge and ophthalmic (Gupta, 1945; Uphof, 1959; Usher, 1974; Bean, 1981; Chopra *et al.*, 1986) menorrhagia, jaundice, chronic diarrhea and piles (Chopra *et al.*, 1986). The plant is used in Skin diseases, diarrhea, diabetes, piles, blood purifier and antispasmodic (Shinwari *et al.,* 2007). |
| | *Folk medicinal use:* | Powder of roots is given for treating diabetes. Decoction of root is given internally for the treatment of pimples and other skin diseases. Roots are used as tooth brush (*Miswak*) for removing toothache. Powder of plant is given with milk to relieve joints pain. The paste of roots is applied on wounds. The aerial branches are boiled in water and given to treat rheumatic fever. The decoction of roots mixed with honey is given to treat jaundice and weakness. |
| | *Family: Cactaceae* | |
| 65 | *Botanical name:* | *Opuntia monacantha* Haw. |
| | *Vernacular name:* | Naghphani |
| | *Part used:* | Fruit, latex and stem |
| | *Flowering period:* | April – May |
| | *Medicinal properties:* | Fruit is known as an expectorant and is useful remedy in asthma, whooping cough; also used as laxative. It is good remedy in hepatic congestion and in gonorrhea. Stem is used to make soothing poultice (Baquar 1989; Nasir and Rafiq 1995; Shinwari and Kan 1998). |
| | *Folk medicinal use:* | The latex is applied on decayed teeth to relieve toothache. |
| | *Family: Euphorbiaceae* | |
| 66 | *Botanical name:* | *Ricinus communis* L. |
| | *Vernacular name:* | Arind/Harnoli |
| | *Part used:* | Oil, leaves and branches |
| | *Flowering period:* | Thought out the year |
| | *Medicinal properties:* | The seed is anthelmintic, cathartic, emollient, laxative, purgative (Grieve, 1984; Chiej, 1984; Lust, 1983). |

| | | |
|---|---|---|
| | | It is rubbed on the temple to treat headache and is also powdered and applied to abscesses and various skin infections. A decoction of the leaves and roots is antitussive, decongestant and expectorant (Duke and Ayensu, 1985). The seed is used in Tibetan medicine, where it is considered to have an acrid, bitter and sweet taste with a heating potency. It is used in the treatment of indigestion and as a purgative (Chopra *et al.*, 1986). The leaves are used as a poultice to relieve headaches and treat boils (Tsarong and Tsewang, 1994). Leprosy, headache, tumor, arthritis, dysentery, impurity of blood and inflammation of the intestine (Srivastava, 1989). |
| | *Folk medicinal use:* | The oil is given to children in constipation. The leaves coated with sesame oil and slightly warmed over fire and applied on painful sites. |
| | *Family: Lamiaceae* | |
| 67 | *Botanical name:* | *Colebrookia oppositifolia* Sm. |
| | *Vernacular name:* | Shakar Dana |
| | *Part used:* | Leaves, branches |
| | *Flowering period:* | January - April |
| | *Medicinal peroperties:* | ---- |
| | *Folk medicinal use:* | The juice of leaves is given to relieve fever, headache and also applied on wounds. |
| | *Family: Solanaceae* | |
| 68 | *Botanical name:* | *Datura stramonium* L. |
| | *Vernacular name:* | Dhatura |
| | *Part used:* | Leaves, seeds |
| | *Flowering period:* | March - August |
| | *Medicinal properties:* | Antispasmodic, asthma, narcotic action and boils (Shinwari *et al.*, 2007). |
| | *Folk medicinal use:* | The powder of seeds is used for cattle worms acting as anthelmintic property. Leaves are warmed over fire and applied on swelled parts to remove pain and inflammation. |
| | *Family: Lamiaceae* | |
| 69 | *Botanical name:* | *Otostegia limbata* (Benth.) Boiss. |
| | *Vernacular name:* | Chitti Bui |

**(Continued)**

| | | |
|---|---|---|
| | *Part used:* | Whole plant |
| | *Flowering period:* | July – November |
| | *Medicinal properties:* | Juice of fresh leaves is taken orally for Gums, scabies remediation. Powder of dried leaves is mixed with butter and layered on wounds and boils. (Haq and Hussain, 1993; Shah and Khan, 2006) |
| | *Folk medicinal use:* | Leaves in combination with leaves of *Olea ferruginea* are chewed to treat mouth infection. |
| | *Family: Malvaceae* | |
| 70 | *Botanical name:* | *Abutilon bidentatum* Hochst. ex Rich. |
| | *Vernacular name:* | Kanghi buti |
| | *Part used:* | Whole plant, leaves |
| | *Flowering period:* | April – October |
| | *Medicinal properties:* | ---- |
| | *Folk medicinal use:* | The powder made up of leaves with few seeds of *Piper nigrum* is given to treat bleeding piles. Leaves powder with *Carum carvi* is given in diarrhea. |
| | *Family: Mimosaceae* | |
| 71 | *Botanical name:* | *Mimosa rubicaulis* Lam. |
| | *Vernacular name:* | Ral |
| | *Part used:* | Leaves and roots |
| | *Flowering period:* | July - August |
| | *Medicinal properties:* | Infusion of leaves is used in piles, while burised leaves are applied on burns. Root powder is used to stop vomiting (Baquar 1989; Nasir and Rafiq 1995; Shinwari and Khan 1998; CSIR 2003; Ali 1973). |
| | *Folk medicinal use:* | Leaves are boiled in water and given to treat vomiting. |
| | *Family: Moraceae* | |
| 72 | *Botanical name:* | *Ficus carica* L. |
| | *Vernacular name:* | Anjeer/Phuwari |
| | *Part used:* | Stem, leaves, fruits |
| | *Flowering period:* | May – November. |

| | | |
|---|---|---|
| | *Medicinal properties:* | The latex from the stems is used to treat corns, warts and piles (Grieve, 1984; Duke and Ayensu, 1985; Polunin and Huxley, 1987). It also has an analgesic effect against insect stings and bites (Chiej, 1984). The fruit is mildly laxative, demulcent, digestive and pectoral (Chiej, 1984; Grieve, 1984; Duke and Ayensu, 1985). The unripe green fruits are cooked with other foods as a galactogogue and tonic (Duke and Ayensu, 1985). The plant has anticancer properties (Duke and Ayensu, 1985). |
| | *Folk medicinal use:* | Fruit is used in constipation and in cough. Fruit is effective in the stomach and gastric problems. . Floral buds are boiled in water and given in abdominal pain, diarrhea and dysentery |
| | *Family: Moraceae* | |
| 73 | *Botanical name:* | *Morus nigra* L. |
| | *Vernacular name:* | Tut |
| | *Part used:* | Fruit, leaves, stem |
| | *Flowering period:* | March - July. |
| | *Medicinal properties:* | The fruit is febrifuge, expectorant, laxative, demulcent, tonic, antibilious and resolvent (Usmanghani *et al.*, 1997). |
| | *Folk medicinal use:* | Bark of stem and leaves are boiled in water and used as gargle for throat pain and toothache. Fruit mixing with potash alum is boiled in water and given as gargle to treat throat pain and inflammation. Plant is supposed to cause allergy during flowering stage. |
| | *Family: Myrsinaceae* | |
| 74 | *Botanical name:* | *Myrsine africana* L. |
| | *Vernacular name:* | Khokhal/Khokhan |
| | *Part used:* | Whole plant, fruit |
| | *Flowering period:* | March - May |
| | *Medicinal properties:* | The fruit is used as an anthelmintic, especially in the treatment of tape worm (Gupta, 1945; Gamble, 1972; Chopra *et al.*, 1986). It is also carminative, stomachic and laxative and is used in the treatment of dropsy and colic (Chopra *et al.*, 1986; Shinwari *et al.*, 2007). A gum obtained from the plant is used as a warming remedy in the treatment of dysmenorrhea. A decoction of the leaf is used as a blood purifier (Chopra *et al.*, 1986). |

**(Continued)**

| | | |
|---|---|---|
| | *Folk medicinal use:* | The powder of fruit is given to expel abdominal worms. Fruits boiled in water and used as gargle to relieve toothache and to kill teeth germs. Leaves are given as fodder to cow to cure dysentery. |
| | *Family: Oleaceae* | |
| 75 | *Botanical name:* | *Nyctanthes arbor-tristis* L. |
| | *Vernacular name:* | Harsingar |
| | *Part used:* | Leaves, flower, bark and seeds |
| | *Flowering period:* | August – October |
| | *Medicinal properties:* | The bark of the plant is expectorant. Leaves are used as febrifuge expectorant and against biliousness; also used in rheumatism; decoction is given in sciatica. Leaf extract is acrid, bitter; useful as cholagogue, laxative, diaphoretic and diuretic, also given to children for the removal of round and thread worms. In unani system of medicine, its flowers are used as astringent to the bowels and in bleeding piles, lessens inflammation and acts as tonic for hair. A decoction of the bark, leaves, roots and flowers is given in excessive diuresis and in enlargement of the spleen. Powdered seeds are used as scurfy affections of the scalp (Baquar 1989; Grohmann 1974; CSIR 2003; NIH 2003). |
| | *Folk medicinal use:* | The paste of flowers is used as hair tonic. Paste of plant is useful for treatment of athlete feet and ringworm. |
| 76 | *Botanical Name:* | *Olea ferruginea* L. |
| | *Vernacular name:* | Kahu |
| | *Part used:* | Leaves, stem |
| | *Flowering period:* | April-May, sometimes September. Fruit: August-November. |
| | *Medicinal properties:* | The decoction of leaves is used in hoarseness of voice and toothache. Leaves and bark are bitter, astringent, antiseptic antiperiodic and diuretic. Fruit is recognized as antidiabetic (Shinwari *et al.*, 2007). |
| | *Folk medicinal use:* | A tea of leaves, sonf (*Foeniculum vulgare*), ajwain (*Trachyspermum ammi*), podina (*Mentha longifolia*) and sugar is taken against cough and flue. Its leaves are rubbed to obtain juice that is used against inflammation. |

| | | |
|---|---|---|
| | *Family: Polygonaceae* | |
| 77 | *Botanical name:* | *Rumex hastatus* D. Don |
| | *Vernacular name:* | Khatimber/Chuki |
| | *Part used:* | Whole plant |
| | *Flowering period:* | April - November |
| | *Medicinal properties:* | The juice of the plant is astringent and is used in the treatment of bloody dysentery. The fresh tuber is chewed to relieve aches in the throat (Manandhar, 2002). |
| | *Folk medicinal use:* | The powder of aerial parts is given for the treatment of diabetes. The powder of flower mixing with table salt is rubbed on teeth to treat tooth problems like pyorrhea and inflammation of gum. |
| | *Family: Punicaceae* | |
| 78 | *Botanical name:* | *Punica granatum* L. |
| | *Vernacular name:* | Druna/Druni |
| | *Part used:* | Fruit, seeds, bark, leaves, branches |
| | *Flowering period:* | August - December |
| | *Medicinal properties:* | The whole plant is antibacterial, antiviral and astringent (Lust, 1959; Parmar and Kaushal, 1982; Lust, 1983; Yeung, 1985; Polunin and Huxley, 1987). The stem bark is emmenagogue. The flowers are used in the treatment of dysentery, stomach ache and cough. The dried pericarp is decocted with other herbs and used in the treatment of colic, dysentery, leucorrhoea etc. (Duke and Ayensu, 1985). Along with the leaves and seeds, they have been used to remove worms. It is used externally in the treatment of vaginal discharges, mouth sores and throat infections. The fruit is a mild astringent and refrigerant in some fevers and especially in biliousness. It is also cardiac and stomachic. The dried rind of the fruit is used in the treatment of amoebic dysentery, diarrhea, etc. The seeds are demulcent and stomachic Both the stem and the root barks are used to expel tapeworms (Grieve, 1984; Chopra *et al.*, 1986). |

**(Continued)**

| | | |
|---|---|---|
| | *Folk medicinal use:* | Fruit has cooling effect used for making *Sharbat*. It is as tonic supposed to reduce iron deficiency. Bark mixed with seeds of *Mangifera indica* is made into powder that is used to treat diarrhea, dysentery. Seeds mixed with *Mentha longifolia* and endocarp of *Juglans regia* are grinded with curd in order to make *Chatni* that is used in stomach problems such as nausea, vomiting, diarrhea, etc. The fruit pericarp is boiled in water to make tea that is given in leucorrhoea. Powder of seeds is used as cooling agent. |
| | *Family: Rhamnaceae* | |
| 79 | *Botanical name:* | *Zizyphus oxyphylla* Edgew. |
| | *Vernacular name:* | Bermaloki |
| | *Part used:* | Fruit |
| | *Flowering period:* | November - December |
| | *Medicinal properties:* | ---- |
| | *Folk medicinal use:* | Fruits are useful in gastric problems. |
| | *Family: Rosaceae* | |
| 80 | *Botanical name:* | *Rosa brunonii* Lindll. |
| | *Vernacular name:* | Jangli gulab |
| | *Part used:* | Flowers and root |
| | *Flowering period:* | March - May |
| | *Medicinal properties :* | The roots are alterative, anodyne, antiphlogistic, antitussive, astringent, diuretic, emmenagogue, expectorant, styptic, tonic and vulnerary (Duke and Ayensu, 1985; Yeung, 1985; Chopra *et al.,* 1986). They are used to lower the blood pressure (Yeung, 1985). The roots are used internally in the treatment of abnormal uterine bleeding, internal and external hemorrhage, bronchitis, rheumatism, stones in the kidney, bladder and gall, dysentery, etc. (Bown, 1985). |

| | | |
|---|---|---|
| | *Folk medicinal use:* | Flowers are crushed mixing with sugar and kept at night under moonlight for few days known as *Gulkand* used to cure constipation. |
| | *Family: Sapindaceae* | |
| 81 | *Botanical name:* | *Dodonaea viscosa* (L.) Jacq. |
| | *Vernacular name:* | Snatha |
| | *Part used:* | Wood |
| | *Flowering period:* | March-July |
| | *Medicinal properties:* | Astringent, rheumatism, swelling and burns (Shinwari *et al.*, 2007). |
| | *Folk Medicinal use:* | Decoction of plant is very effective in sugar diseases (Diabetes). Ash of leaves grinded with mustard oil is applied externally on boils. For rheumatic pain, leaves are coated with oil and slightly warmed on fire and bound. |
| | *Family: Solanaceae* | |
| 82 | *Botanical name:* | *Solanum incanum* L. |
| | *Vernacular name:* | Mahori |
| | *Part used:* | Leaves, aerial parts and root |
| | *Flowering period:* | Throughout the year |
| | *Medicinal properties:* | Root is used as a medicine for horses. Plant used in toothache and sore throat; decoction is used in chest troubles (Baquar 1989; Nasir 1985). |
| | *Folk medicinal use:* | Aerial parts mixing with the leaves of *Adhatoda zeylania* and fruits of *Vitis vinifera* are boiled in water and given to treat flue, cough and fever. |
| 83 | *Botanical name:* | *Solanum nigrum* L. |
| | *Vernacular name:* | Peelan/Kachmach |

**(Continued)**

| | | |
|---|---|---|
| | *Part used:* | Whole plant, leaves, stem and root |
| | *Flowering period:* | March - May |
| | *Medicinal properties:* | The whole plant is antiperiodic, antiphlogistic, diaphoretic, diuretic, emollient, febrifuge, narcotic, purgative and sedative (Singh and Kachroo, 1976; Emboden, 1979, Lust, 1983; Grieve, 1984, Duke and Ayensu, 1985). It is harvested in the autumn when both flowers and fruit are upon the plant, and is dried for later use (Grieve, 1984). Use with caution (Lust, 1983). The leaves, stems and roots are used externally as a poultice; wash etc in the treatment of cancerous sores, boils, leucoderma and wounds (Duke and Ayensu, 1985, Moerman, 1998). Extracts of the plant are analgesic, antispasmodic, anti-inflammatory and vasodilator (Duke and Ayensu, 1985). The plant has been used in the manufacture of locally analgesic ointments and the juice of the fruit has been used as an analgesic for toothaches (Chiej, 1984).Diuretic, expectorant, rheumatic, gonorrhea, heart diseases and skin diseases (Shinwari *et al.,* 2007). |
| | *Folk medicinal use:* | The decoction of plant is used for digestive problems. Juice of the leaves is given in hiccough and inflammation of eye. Leaves are cooked as vegetable to remove swelling, dropsy and digestive problems. Leaves are soaked in water and are kept overnight. These leaves are squeezed and extract is used against jaundice. The poultice of leaves is applied on to remove pain and inflammation. |
| 84 | *Botanical name:* | *Solanum surattense* Burm.f. |
| | *Vernacular name:* | Kandiali |
| | *Part used:* | Root and fruits |
| | *Flowering period:* | March – May |
| | *Medicinal properties:* | Diuretic, expectorant, rheumatic, gonorrhea, heart diseases and skin diseases (Shinwari *et al.,* 2007). |

| | | |
|---|---|---|
| | *Folk medicinal use:* | The roots mixed with *Piper longum* and *P. nigrum* is boiled in water and honey is added. This formula is given to treat flue, cough and asthma. Fruit powder is mixed with honey and is given in cough, asthma and joints pain. The ash of plant mixed with honey is given to treat cough and asthma. |
| 85 | *Botanical name:* | *Withania somnifera* (L.) Dunal. |
| | *Vernacular name:* | Asghand/Aksan |
| | *Part used:* | Roots and leaves |
| | *Flowering period:* | Throughout the year |
| | *Medicinal properties:* | The plant is little known in the West (Emboden, 1979). The whole plant, but especially the leaves and the root bark, are abortifacient, adaptogen, antibiotic, aphrodisiac, deobstruent, diuretic, narcotic, strongly sedative and tonic (Emboden, 1979; Chopra *et al.*, 1986; Bown, 1995). Internally, it is used to tone the uterus after a miscarriage and also in treating post-partum difficulties (Emboden, 1979). It is also used to treat nervous exhaustion, debility, insomnia, wasting diseases, failure to thrive in children, impotence, infertility, multiple sclerosis etc (Bown, 1995). Externally it has been applied as a poultice to boils, swellings and other painful parts (Emboden, 1979, Chopra *et al.*, 1986). The root is harvested in the autumn and dried for later use (Bown, 1995). Some caution is advised in the use of this plant since it is toxic (K). The fruit is diuretic. The seed is diuretic and hypnotic (Chopra *et al.*, 1986). Rheumatism, lumber pain, swelling, bronchitis, tumours and tubercular glands (Shinwari *et al.,* 2007). |
| | *Folk Medicinal use:* | The powder of leaves is used against inflammation. Fumes (*Hukka*) of the roots used for toothache and headache. Roots are powdered and taken with milk for nervous system. |
| | *Family: Urticaceae* | |
| 86 | *Botanical name:* | *Debregeasia salicifolia* (D. Don) Rendle |

**(Continued)**

| | | |
|---|---|---|
| | *Vernacular name:* | Sinduri/Sindur |
| | *Part used:* | Fruit |
| | *Flowering period:* | July- September |
| | *Medicinal properties:* | The decoction of leaves with herbs used in diabetic (Shinwari *et al.,* 2007). |
| | *Folk Medicinal use:* | Fruit is edible used as tonic. People think that the fruit has some narcotic effects. |
| | *Family: Verbenaceae* | |
| 87 | *Botanical name:* | *Vitex negundo* L. |
| | *Vernacular name:* | Banna |
| | *Part used:* | Leaves, stem, roots and seeds. |
| | *Flowering period:* | Whole year |
| | *Medicinal properties:* | Root is tonic, febrifuge, anthelmintic, demulcent and expectorant; used in dyspepsia, rheumatism, piles and boils in powdered form. Leaves are aromatic, tonic and vermifuge; used in the form of decoction in catarrhal fever and swelling of joints. Fresh leaves are used in the form of poultice for painful chest and back; also used in gum and skin diseases. Dried leaves are smoked in headache. Extract of the leaves show anticancer properties and is employed as hair tonic. Dried fruit is vermifuge. In unani system of medicine its seeds are employed in inflammations, flatulence and reduce the quality of semen (Baquar 1989; Nasir and Rafiq 1995; Shanwari and Khan 1998; CSIR 2003, NIH 2003; Jafir and Ghafoor 1974). |
| | *Folk medicinal use:* | Leaves are grinded with black pepper and made into pills which are given to treat piles. |
| | *Family: Violaceae* | |
| 88 | *Botanical name:* | *Woodfordia fruticosa* (L.) S.Kurz |
| | *Vernacular name:* | Tavi |

| | |
|---|---|
| *Part used:* | Flowers |
| *Flowering period:* | March - April |
| *Medicinal properties:* | Menorrhagia, hemorrhoids, astringent, antipyretic and backache (Shinwari *et al.,* 2007). |
| *Folk Medicinal use:* | Flowers are dried and made into powder and given in leucorrhoea, menorrhagia and piles. The decoction of flowers is given in diarrhea. |

## 4.1.9. Trees

| | *Family: Anacardiaceae* | |
|---|---|---|
| 89 | *Botanical name:* | *Lannea coromandelica* (Houtt.) Merr. |
| | *Vernacular name:* | Kamlai |
| | *Part used:* | Bark |
| | *Flowering period:* | February – may |
| | *Medicinal properties:* | ---- |
| | *Folk medicinal use:* | Decoction of bark is used in mouth ulcer (Aphthae). |
| 90 | *Botanical name:* | *Pistacia chinensis* Bunge |
| | *Vernacular name:* | Kangar |
| | *Part used:* | Fruit. |
| | *Flowering period:* | March - April |
| | *Medicinal properties:* | Galls are aromatic, tonic and expectorant used in phthisis, cough and asthma. Powdered galls fried with ghee are given internally in dysentery. It is also an antidote to snake venom and scorpion sting. The oil from the galls is used as carminative. In moderate doses, it possesses an antispasmodic action on involuntary muscles, inhibiting excessive peristaltic movements of intestines (Baquar 1989; Nasir and Rafiq 1995 and Shinwari and Khan 1998; CSIR 2003; Nasir 1983). |
| | *Folk medicinal use:* | The powder of fruit mixed with honey is given to treat cough, cold and asthma. |

**(Continued)**

| | | |
|---|---|---|
| | *Family: Bombacaceae* | |
| 91 | *Botanical name:* | *Bombax ceiba* L. |
| | *Vernacular name:* | Simbal |
| | *Part used:* | Wood, root, fruit, fluff of capsule, seed and leaves. |
| | *Flowering period:* | December - March |
| | *Medicinal properties:* | Young roots used as astringent, alternative, restorative and as brain tonic and treatment for gonorrhea and dysentery. Gum is astringent used to treat diarrhea and menorrhagia. Leaves powder is mixed in milk and used in stangury. Young fruits are used as stimulant and aphrodisiac. Second are used in gonorrhea, gleet, chronic cystitis, pulmonary tuberculosis and catarrhal fever (Nasir and Rafiq 1995; Sheikh 1993; Shinwari and Khan 1998; Polunin and Stainton 1987; Qaiser 1978; NIH, 2003). |
| | *Folk medicinal use:* | The powder of gum is given to treat diarrhea. |
| | *Family: Ebenaceae* | |
| 92 | *Botanical name:* | *Diospyros lotus* L. |
| | *Vernacular name:* | Amlok |
| | *Part used:* | fruit |
| | *Flowering period:* | November- December |
| | *Medicinal properties:* | ---- |
| | *Folk medicinal use:* | Fruit is used in stomach problems. |
| 93 | *Botanical name:* | *Mallotus philipensis* (Lam.) Muell. |
| | *Vernacular name:* | Kamela |
| | *Part used:* | Fruits |
| | *Flowering period:* | February - March |
| | *Medicinal properties:* | ---- |
| | *Folk medicinal use:* | The epicarpic powder of fruits is mixed with curd and given to expel abdominal worms. |
| | *Family: Euphorbiaceae* | |
| 94 | *Botanical name:* | *Phyllanthu emblica* L. |
| | *Vernacular name:* | Amla |
| | *Part used:* | Root, bark and fruit |

| | | |
|---|---|---|
| | *Flowering period:* | March – May |
| | *Medicinal properties:* | Root and root bark is astringent. Leaves and barks are used as poultice in cutaneous diseases. Flowers are used as refrigerant, cooling and aperient. Exudation from the fruit is externally applied to inflamed eyes. Powder obtained from the fruits is used as a vermifuge, purgative and in certain parasitic skin diseases; also employed in hemorrhage, diarrhea, dysentery, jaundice, dyspepsia and anemia. Seeds are uses in asthma, bronchitis biliousness. Fruit used in pickle, preserves and jellies and are probably the richest source of vitamin C (Baquar, 1989; Smith, 1986; Sheikh 1993; Nasir and Rafiq, 1995; Shinwari and Khan, 1998; NIH 2003). |
| | *Folk medicinal use:* | The fruit is made into murabba, a traditional preserving art in sugar solution which is used to strengthen stomach. |
| | *Family: Fabaceae* | |
| 95 | *Botanical name:* | *Pongamia pinnata* (L.) Pierre |
| | *Vernacular name:* | Sukh chain |
| | *Part used:* | Whole plant |
| | *Flowering period:* | April-May |
| | *Medicinal properties:* | Bark is recommended internally in bleeding piles; juice of rood is used against fistulous sores and for cleaning foul ulcers; given intensity with equal amount of coconut milk and lime in gonorrhea. Juice of leaves is prescribed in flatulence, dyspepsia, diarrhea and cough; also considered as a remedy for leprosy and gonorrhea. A hot infusion of leaves is used as a medicated bath for rheumatic joints and for cleaning foul ulcers and sores. Dried flowers are used in the form of decoction to quench thirst in diabetes. Oil from seeds in useful in cutaneous affections, herpes and scabies, mixed with lime and lemon juice is useful in the treatment of rheumatism (Baquar 1989; Ali 1977; CSIR 2003). |

**(Continued)**

| | | |
|---|---|---|
| | *Folk medicinal use:* | Twigs are used as tooth stick for strengthening of gums. |
| | *Family: Fagaceae* | |
| 96 | *Botanical name:* | *Quercus leucotrichophora* A. Camus |
| | *Vernacular name:* | Shah-e-baloot |
| | *Part used:* | Acorn |
| | *Flowering period:* | April-May |
| | *Medicinal properties:* | Acorn is used as an astringent and diuretic in diarrhea, indigestion asthma and gonorrhea (Nasir 1976; Baquar 1989; Nasir and Rafiq 1995; Shinwari and Khan 1998; CSIR 2003; NIH 2003). |
| | *Folk medicinal use:* | Acorn is burnt in fire and dusted on wounds to heal. |
| | *Family: Flacourtiaceae* | |
| 97 | *Botanical name:* | *Flacourtia indica* (Burm. f.) Merrill |
| | *Vernacular name:* | Kokoh |
| | *Part used:* | Fruit |
| | *Flowering period:* | March - April. |
| | *Medicinal properties:* | ---- |
| | *Folk medicinal use:* | Ripened fruit is given in diabetes. The same is used as tonic. |
| | *Family: Meliacea* | |
| 98 | *Botanical name:* | *Melia azedarach* L. |
| | *Vernacular name:* | Dharek |
| | *Part used:* | Leaves |
| | *Flowering period:* | March - April. |

| | | |
|---|---|---|
| | *Medicinal properties:* | Anthelmintic, diuretic, vomiting, leucoderma, rheumatism and blood purifier (Lust, 1983; Grieve, 1984; Duke and Ayensu; 1985; Yeung, 1985; Chopra *et al.*, 1986; Bown, *1995*). |
| | *Folk medicinal use:* | Plant supposed to have cooling effect and decoction of leaves is given in pimples and boils. The decoction is also given in diabetic patients. Leaves are given as fodder to cows in the case of fever. |
| 99 | *Botanical name:* | *Acacia modesta* Wall. |
| | *Part used:* | Bark, stem, leaves and gum |
| | *Flowering period:* | March - May. |
| | *Medicinal properties:* | Gum is used as tonic and stimulant given to women in child birth (Shinwari *et al.*, 2007). |
| | *Folk medicinal use:* | Powder of stem bark mixing with salt is used as gargle to relieve in toothache. Twigs of *Fagonia indica, Swertia chiratia, Acacia nilotica* and *Acacia modesta* are boiled in about half of water in a big pan until a bottle of water is left. This extract is sieved and stored in a bottle and 3-4 spoons of this extract used daily in the morning not only by the patients but also by common peoples in summers in jaundice and as cooling agent. Halwa (A cooked dessert of Gum + Desi ghee + Poppy seeds + Eggs + Fennel + Almond + Rattan jo + Pistachio) is fed to women after child birth. For abdominal pain decoction of bark is taken |
| 100 | *Botanical name:* | *Albizzia lebbeck* (L.) Benth. |
| | *Vernacular name:* | Shirin |
| | *Part used:* | Leaves, wood |
| | *Flowering period:* | March - May |
| | *Medicinal properties:* | ---- |
| | *Folk medicinal use:* | The powder of bark is dusted on wounds to heal. The bark is boiled in water and used as gargle for strengthening of gums and teeth. |

**(Continued)**

| | | |
|---|---|---|
| | *Family: Moraceae* | |
| 101 | *Botanical name:* | *Ficus bengalensis* L. |
| | *Vernacular name:* | Barr |
| | *Part used:* | Leaves, fruit and latex |
| | *Flowering period:* | April – May |
| | *Medicinal properties:* | Latex is externally applied for pains and bruises and for the treatment of cracked or inflamed soles of feet; as an anodne in rheumatism and lumbago; as a remedy in toothache; also used as an aphrodisiac. The leaves are heated as a poultice, applied to abscesses. The root fibers are said to be given in gonorrhea. Seeds are used as cooling and tonic (Baquar, 1989; Ghafoor 1985; Shinwari and Khan 1998; CSIR 2003; NIH 2003). |
| | *Folk medicinal use:* | Latex is poured on sugar (*Batasha*) and orally given as sexual tonic. Fruits are dried and made into powder and used as tonic. |
| 102 | *Botanical name:* | *Ficus roxburghii* Wall. ex Brand. |
| | *Vernacular name:* | Dusi |
| | *Part used:* | Fruits |
| | *Flowering period:* | August - November. |
| | *Medicinal properties:* | Laxative, antidiabetic, piles, constipation and stomach problems. Fruits are used as cooling, appetizer and tonic. (Srivastava, 1989). |
| | *Folk medicinal use:* | Fruit are eaten to treat constipation and piles. Unripe fruits are cooked as vegetable for diarrhea and to stop bleeding piles. Milky latex is applied on boils. |
| | *Family: Moraceae* | |
| 103 | *Botanical name:* | *Morus alba* L. |

| | | |
|---|---|---|
| | *Vernacular name:* | Shehtoot |
| | *Part used:* | Fruit |
| | *Flowering period:* | April - September |
| | *Medicinal properties:* | Fruit is anthelmintic, astringent, purgative, diaphoretic, emollient and laxative (Shinwari *et al.,* 2007). |
| | *Folk medicinal use:* | Dried fruits are given in constipation to treat. During winter season, the dried fruits are used as tonic for energy source. |
| | *Family: Oleaceae* | |
| 104 | *Botanical name:* | *Olea europaea* L. |
| | *Vernacular name:* | Koh |
| | *Part used:* | Oil, leaves, fruits |
| | *Flowering period:* | April - May. Fruit: September - October. |
| | *Medicinal properties:* | The oil is cholagogue, nourishing demulcent, emollient and laxative (Lust, 1983; Grieve, 1984; Chopra *et al.,* 1986). Eating the oil reduces gastric secretions and is therefore of benefit to patients suffering from hyperacidity (Grieve, 1984; Bown, 1995). It is used externally to treat pruritis, the effects of stings or burns and as a vehicle for liniments. Used with alcohol it is a good hair tonic and used with oil of rosemary it is a good treatment for dandruff. The leaves are antiseptic, astringent, febrifuge and sedative (Lust, 1983; Grieve, 1984). A decoction is used in treating obstinate fevers; they also have a tranquillizing effect on nervous tension and hypertension (Grieve, 1984; Bown, 1995). The bark is astringent, bitter and febrifuge (Grieve, 1984; Chopra *et al.*, 1986). It is said to be a substitute for quinine in the treatment of malaria (Chopra *et al.*, 1986). |

**(Continued)**

| | | |
|---|---|---|
| | *Folk Medicinal use:* | Oil is applied on hairs for the removal of dandruff. Fruit of plant is used in diabetes. |
| | *Family: Pinaceae* | |
| 105 | *Botanical name:* | *Pinus roxburghii* Sargent |
| | *Vernacular name:* | Chir |
| | *Part used:* | Resin and leaves, whole plant |
| | *Flowering period:* | Almost throughout the year |
| | *Medicinal properties:* | The turpentine obtained from the resin of all pine trees is antiseptic, diuretic, rubifacient, antirheumatic and vermifuge. It is also very beneficial to the respiratory system and so is useful in treating diseases of the mucous membranes and respiratory complaints such as coughs, colds, influenza and TB. Externally it is a very beneficial treatment for a variety of skin complaints, wounds, sores, burns, boils etc and is used in the form of liniment plasters, poultices, herbal steam baths and inhalers (Grieve, 1984; Usmanghani*et al.*, 1997). The wood is diaphoretic and stimulant. It is useful in treating burning of the body, cough, fainting and ulcers (Chopra *et al.*, 1986). |
| | *Folk medicinal use:* | Patients suffering from Tuberculosis are advised to keep sitting under the shade of this plant to heal. Juvenile leaves are crushed with milk and obtained juice is given in chicken pox (*Khasra*). Fresh sap (Resin) is applied as plaster on the boils to hasten and remove pus. |
| | *Family: Rhamnaceae* | |
| 107 | *Botanical name:* | *Zizyphus mauritiana* Mill. |
| | *Vernacular name:* | Beri |
| | *Part used:* | Fruit and leaves |
| | *Flowering period:* | July - September |

| | | |
|---|---|---|
| | *Medicinal properties:* | Febrifuge, anthelmintic, ulcer, astringent, blood purifier and anodyne (Shinwari *et al.,* 2007). |
| | *Folk medicinal use:* | Fruits are used to cure constipation. Leaves are boiled in water and used as bath to treat dandruff and for healthy and shiny hairs. |
| | *Family: Rosaceae* | |
| 108 | *Botanical name:* | *Pyrus pashia* Ham ex D. Don |
| | *Vernacular name:* | Batangi |
| | *Part used:* | Fruit |
| | *Flowering period:* | March - April |
| | *Medininal properties:* | Fruit is used as febrifuge, sedative and astringent (Nasir and Rafiq 1995; Sheikh 1993; Shinwari and Khan 1998; CSIR 2003). |
| | *Folk Medicinal use:* | Fruit is eaten as tonic. |
| | *Family: Tiliaceae* | |
| 109 | *Botanical name:* | *Grewia optiva* Drum. ex Burret. |
| | *Vernacular name:* | Taman |
| | *Part used:* | Fruits |
| | *Flowering period:* | March – October |
| | *Medicinal properties:* | ---- |
| | *Folk medicinal use:* | Powder of fruit is used for dysentery |

**Table 4. Novel medicinal uses of plants recorded from Kotli Sattian**

| S. No. | Plant species | Medicinal uses |
|---|---|---|
| 1 | *Acacia modesta* Wall. | Toothache, jaundice and as cooling agent |
| 2 | *Abutilon bidentatum* Hochst. ex Rich. | Bleeding piles and diarrhea. |
| 3 | *Achyranthes aspera* L. | Toothache, hasten delivery. |
| 4 | *Adhatoda zeylanica* Medic. | Dysentery, diabetic, jaundice, boils |

**Table 4. (Continued)**

| S. No. | Plant species | Medicinal uses |
|---|---|---|
| 5 | *Ajuga bracteosa* Wall. ex Bth. | Constipation and piles, cooling agent. |
| 6 | *Albizzia lebbeck* (L.) Benth. | Wounds and pyorrhea. |
| 7 | *Amaranthus hybridus* L. | Constipation. |
| 8 | *Amaranthus viridis* L. | Constipation. |
| 9 | *Bacopa monnieri* (L.) Wettst. | Skin irritation and urticaria. |
| 10 | *Berberis lyceum* Royle. | Pimples, toothache, joints pain and wounds. |
| 11 | *Bombax ceiba* L. | Diarrhea. |
| 12 | *Calotropis procera* (Willd.) R. Br. | Boils and rheumatic pain. |
| 13 | *Carissa opaca* Stapf ex. Haines | Hepatitis, flue and headache, diabetes and jaundice. |
| 14 | *Chenopodiumam ambrosioides* L. | Headache. |
| 15 | *Colchicum aitchisonii* (Hook. f.) E. Nasir | Wounds. |
| 16 | *Colebrookia oppositifolia* Sm. | Fever, headache and wounds. |
| 17 | *Cuscuta reflexa* Roxb. | Chronic fever, hepatitis. |
| 18 | *Cynodon dactylon* (L.) Pers. | Wounds |
| 19 | *Cyperus rotundus* L. | Cholera. |
| 20 | *Datura stramonium* L. | Anthelmintic, pain and inflammation. |
| 21 | *Debregeasia salicifolia* (D. Don) Rendle | Tonic and narcotic. |
| 22 | *Diospyros lotus* L. | Stomach problems. |
| 23 | *Dodonaea viscosa* (L.) Jacq. | Diabetes and boils. |
| 24 | *Eruca sativa* L. | Pimples, leprosy and skin dandruff. |
| 25 | *Ficus bengalensis* L. | Aphrodisiac. |
| 26 | *Ficus carica* L. | Cough, abdominal pain, diarrhea and dysentery. |
| 27 | *Ficus roxburghii* Wall. ex Brand. | Diarrhea and boils. |
| 28 | *Flacourtia indica* (Burm. f.) Merrill | Diabetes and tonic. |
| 29 | *Grewia optiva* Drum. ex Burret. | Dysentery. |
| 30 | *Hedera nepalensis* K. Koch | Wounds and boils. |
| 31 | *Heliotropium europaeum* L. | Skin irritation. |
| 32 | *Ipomoea carnea* Jacq. | Pain and inflammation. |

| S. No. | Plant species | Medicinal uses |
|---|---|---|
| 33 | *Lannea coromandelica* (Houtt.) Merr. | Mouth ulcer (Aphthae). |
| 34 | *Launaea procumbens* (Roxb.) Ram. & Rajgo. | Skin rashes. |
| 35 | *Mallotus philipensis* (Lam.) Muell. | Anthelmintic (abdominal worms). |
| 36 | *Melia azedarach* L. | Diabetes. |
| 37 | *Morus nigra* L. | Throat pain and inflammation. |
| 38 | *Nyctanthes arbor-tristis* L. | Hair tonic, athlete feet and ringworm. |
| 39 | *Olea europaea* L. | Antidandruff and anti-diabetic. |
| 40 | *Olea ferruginea* L. | Flue, cough and inflammation. |
| 41 | *Origanum vaulgare* L. | Appetizer. |
| 42 | *Oxalis corniculata* L. | Dysentery and jaundice. |
| 43 | *Peganum harmala* L. | Pimples, boils and carbuncle, small pox (*Khasra*). |
| 44 | *Periploca aphylla* Decne. | Boils. |
| 45 | *Phyla nodiflora* (L.) Greene | Bleeding piles. |
| 46 | *Physalis divaricata* D. Don. | Laxative. |
| 47 | *Pinus roxburghii* Sargent | Chicken pox (*Khasra*), wounds. |
| 48 | *Plantago ovata* Frossk. | Obesity. |
| 49 | *Polygonum plebejum* R. Br. | Fever, throat pain and cooling agent. |
| 50 | *Pongamia pinnata* (L.) Pierre | Pyorrhea. |
| 51 | *Pueraria tuberosa* (Roxb. Ex Willd) DC. | Joints pain and inflammation. |
| 52 | *Punica granatum* L. | Cooling agent, tonic, nausea, vomiting, diarrhea, leucorrhoea. |
| 53 | *Pyrus pashia* Ham ex D. Don | Tonic. |
| 54 | *Quercus leucotrichophora* A. Camus | Wounds healing. |
| 55 | *Ranunculus arvensis* L. | Ringworm and athlete foot. |
| 56 | *Ranunculus muricatus* L. | Ringworm and athlete foot. |
| 57 | *Rosa brunonii* Lindll. | Constipation. |
| 58 | *Rubia cordifolia* L. | Emmenogogue (regulating menstrual period). |
| 59 | *Rubus ellipticus* Smith | Tonic. |
| 60 | *Rumex dentatus* L. | Constipation and piles. |
| 61 | *Rumex hastatus* D. Don | Diabetes and pyorrhea. |

**Table 4. (Continued)**

| S. No. | Plant species | Medicinal uses |
|---|---|---|
| 62 | *Rumex nepalensis* Spreng. | Anti-poisonous. |
| 63 | *Salvia moorcroftiana* Wall. ex Bth. | Inflamed parts. |
| 64 | *Sida cordata* (Burm. f.) Borss.-Waalkes | Joints pain. |
| 65 | *Sisymbrium irio* L. | Chronic fever. |
| 66 | *Smilax aspera* L. | Wounds |
| 67 | *Solanum incanum* L. | Flue, cough and fever. |
| 68 | *Solanum nigrum* L. | Hiccough, jaundice, pain and inflammation. |
| 69 | *Tagetes minuta* L. | Earache. |
| 70 | *Tribulus terrestris* L. | Constipation, child birth delivery, leucorrhoea and back ache. |
| 71 | *Trichodesma indicum* (L.) R. Br. | Diarrhea (bloody motions), abdominal pain and as cooling agent. |
| 72 | *Valeriana wallichii* DC. | Diarrhea. |
| 73 | *Verbascum thapsus* L. | Migraine, pain and sleeplessness. |
| 74 | *Verbena officinalis* L. | Rheumatic and joints pain. |
| 75 | *Viola cancescens* Wall. ex Roxb. | Throat pain, flue, cough and asthma. |
| 76 | *Withania somnifera* (L.) Dunal. | Toothache and headache. |
| 77 | *Woodfordia fruticosa* (L.) S.Kurz | Leucorrhoea and diarrhea. |
| 78 | *Zanthoxylum alatum* Roxb. | Piles and constipation, nausea, vomiting and abdominal pain. |
| 79 | *Zizyphus mauritiana* Mill. | Constipation and antidandruff. |
| 80 | *Zizyphus oxyphylla* Edgew. | Gastric problems. |

**Table 5. Use of various plant parts for making anecdotal recipes**

| S. No. | Parts used | No. of species | percentage |
|---|---|---|---|
| 1 | Leaves | 57 | 29.08 |
| 2 | Whole plant | 35 | 17.86 |
| 3 | Fruit | 26 | 13.27 |
| 4 | Stems | 23 | 11.73 |
| 5 | Roots | 22 | 11.22 |
| 6 | Seeds | 16 | 8.16 |
| 7 | Flower | 9 | 4.59 |
| 8 | Gum/latex | 6 | 3.06 |
| 9 | Oil | 2 | 1.02 |
| | | 196 | 100.00 |

**Table 6. Different forms of medicaments for the treatment of various diseases**

| S. No. | Formulations | No. of species | Percentage |
|---|---|---|---|
| 1 | Decoction | 29 | 20.28 |
| 2 | Powder | 28 | 19.58 |
| 3 | Raw form | 22 | 15.38 |
| 4 | Juice | 16 | 11.19 |
| 5 | Extract | 13 | 9.09 |
| 6 | Paste | 11 | 7.69 |
| 7 | Potherb | 9 | 6.29 |
| 8 | Poultice | 6 | 4.20 |
| 9 | Fumes | 6 | 4.20 |
| 10 | Husk | 3 | 2.10 |
| | | 143 | 100.00 |

## 4.1.10. Novel Medicinal Use

Comparing with the medicinal literature, seven species viz., *Abutilon bidentatum, Colebrookia oppositifolia, Ipomoea carnea, Trichodesma indicum, Valeriana wallichii, Verbena officinalis* and *Rubus ellipticus* possessed novel uses not being recorded before. Futhermore, 73 species had additional uses. The list of species along with their uses is provided below in Table 4. Some species such as *Adhatoda zeylanica, Berberis lyceum, Carissa opaca, Ficus carica, Punica granatum* and *Zanthoxylum alatum* are heavily used by the natives for treating about four different diseases. These species are very commonly distributed in the area under study.

## 4.1.11. Parts Used

Table 5 revealed that leaf was the most preferred parts of the plant for making crude medicine (57 spp., 30.98%), followed by whole plant (35 spp., 19.02%), fruit (26 spp., 14.13%), stems (23 spp., 12.50%), roots (22 spp., 11.96%), seeds (16 spp., 8.70%) and flowers (9 spp., 4.89%), while rest of parts were less used.

**Table 7. Inventory of palatable flora along with family, parts used, palatability, occurence and animal preference**

| S. No. | Plant Name | Family | Parts used | | | Palatability | | | Occurence | | | Animal preference |
|---|---|---|---|---|---|---|---|---|---|---|---|---|
| | | | WP | AP | Lv | LP | MP | HP | C | VC | R | |
| 1 | *Abutilon bidentatum* Hochst. ex Rich. | Malvaceae | X | √ | X | X | √ | X | X | X | √ | Goat |
| 2 | *Arabis himalaica* (Edgew.) O.E. Schulz | Brassicaceae | X | √ | X | X | X | √ | X | X | √ | Goat, Cow |
| 3 | *Acacia modesta* Wall. | Mimosaceae | X | X | √ | X | X | √ | √ | X | X | Goat, Sheep |
| 4 | *Achyranthus aspera* L. | Amaranthaceae | √ | X | X | X | X | √ | X | √ | X | Goat, Sheep, Cow |
| 5 | *Adiantum capillus-veneris* L. | Adiantaceae | √ | X | X | X | X | √ | X | √ | X | Goat |
| 6 | *Adiantum incissum* L. | Adiantaceae | √ | X | X | X | X | √ | X | √ | X | Goat |
| 7 | *Ailanthus altissima* (Mill.) Swingle | Simarubaceae | X | X | X | √ | X | X | √ | X | X | Goat |
| 8 | *Ajuga bracteosa* Wall. ex Bth. | Lamiaceae | √ | X | X | √ | X | X | √ | X | X | Goat, Sheep |
| 9 | *Ajuga parviflora* Bth. | Lamiaceae | √ | X | X | √ | X | X | √ | X | X | Goat, Sheep |
| 10 | *Albizzia lebbeck* (L.) Benth. | Mimosaceae | X | X | √ | √ | X | X | X | X | √ | All |
| 11 | *Alternanthera pungens* Kunth. | Amaranthaceae | X | X | √ | X | √ | X | X | X | √ | Goat, Cow, Donkey |
| 12 | *Amaranthus hybridus* L. | Amaranthaceae | √ | X | X | X | X | √ | √ | X | X | Goat, Cow, Donkey |
| 13 | *Amaranthus spinosus* L. | Amaranthaceae | √ | X | X | X | X | √ | √ | X | X | All |

| S. No. | Plant Name | Family | Parts used | | | Palatability | | | Occurence | | | Animal preference |
|---|---|---|---|---|---|---|---|---|---|---|---|---|
| | | | WP | AP | Lv | LP | MP | HP | C | VC | R | |
| 14 | *Amaranthus viridis* L. | Amaranthaceae | √ | X | X | X | X | √ | X | √ | X | All |
| 15 | *Anagalis arvensis* L. | Primulaceae | √ | X | X | X | X | √ | √ | X | X | All |
| 16 | *Argyrolobium helleborifolium* Sonott. | Fabaceae | √ | X | X | X | X | √ | X | X | √ | All |
| 17 | *Argyrolobium roseum* (Camb.) Jaub & Spach | Fabaceae | X | √ | X | X | X | √ | X | X | √ | All |
| 18 | *Aristida cyanatha* Nees ex Steud. | Poaceae | X | √ | X | X | √ | X | √ | X | X | All |
| 19 | *Artemisia scoparia* Waldst. & Kit. | Asparagaceae | X | X | √ | √ | X | X | √ | X | X | Donkey |
| 20 | *Arundo donax* L. | Poaceae | X | X | √ | X | √ | X | X | X | √ | Goat, Sheep, Donkey |
| 21 | *Asphodelus tenuifolius* Cavan. | Liliaceae | X | X | √ | X | X | √ | √ | X | X | All |
| 22 | *Astragalus squarrosus* Bunge | Fabaceae | X | X | √ | X | √ | X | X | X | √ | Goat, Sheep |
| 23 | *Avena fatua* L. | Poaceae | X | X | √ | X | X | √ | √ | X | X | All |
| 24 | *Bacopa monnieri* (L.) Wettst. | Scrophulariaceae | √ | X | X | X | √ | X | √ | X | X | Goat, Sheep |
| 25 | *Barleria acanthoides* Vahl | Acanthaceae | X | √ | X | X | √ | X | X | X | √ | All |
| 26 | *Barleria cristata* L. | Acanthaceae | X | X | √ | X | √ | X | X | X | √ | All |
| 27 | *Berberis lycium* Royle. | Berberidaceae | X | X | √ | √ | X | X | X | √ | X | Goat |

**Table 7. (Continued)**

| S. No. | Plant Name | Family | Parts used | | | Palatability | | | Occurence | | | Animal preference |
|---|---|---|---|---|---|---|---|---|---|---|---|---|
| | | | WP | AP | Lv | LP | MP | HP | C | VC | R | |
| 28 | *Boerhavia procumbens* Banks ex Roxb. | Nyctaginaceae | √ | X | X | X | X | √ | X | √ | X | All |
| 29 | *Bombax ceiba* L. | Bombacaceae | X | X | √ | X | √ | X | √ | X | X | All |
| 30 | *Calotropis procera* (Willd.) R.Br. | Asclepiadaceae | X | X | √ | √ | X | X | X | X | √ | Goat |
| 31 | *Commelina paludosa* Blume | Commelinaceae | X | √ | X | X | X | X | √ | X | X | Goat, Sheep, Cow |
| 32 | *Cannabis sativa* L. | Cannabinaceae | X | √ | X | √ | X | X | X | √ | X | Goat |
| 33 | *Capsella bursa-pastoris* (L.) Medik | Brassicaceae | √ | X | X | X | X | √ | X | √ | X | All |
| 34 | *Cardiospermum halicacabum* L. | Sapindaceae | X | √ | X | X | √ | X | X | X | √ | Goat |
| 35 | *Carissa opaca* Stapf ex. Haines | Apocynaceae | X | X | √ | √ | X | X | √ | X | X | Goat |
| 36 | *Carthamus oxycantha* M. Bieb | Asteraceae | X | √ | X | X | √ | X | X | X | √ | All |
| 37 | *Cenchrus ciliaris* L. | Poaceae | X | √ | X | X | X | √ | √ | X | X | All |
| 38 | *Cenchrus pennisetiformis* Hochst. & Steud. | Poaceae | √ | X | X | X | X | √ | √ | X | X | All |
| 39 | *Cenchrus setigerus* Vahl | Poaceae | X | √ | √ | X | X | √ | √ | X | X | All |
| 40 | *Chenopodium album* L. | Chenopodiaceae | X | X | X | X | X | √ | √ | X | X | All |
| 41 | *Chenopodium ambrosioides* L. | Chenopodiaceae | X | X | √ | X | X | √ | √ | | | Goat |

| S. No. | Plant Name | Family | Parts used | | | Palatability | | | Occurence | | | Animal preference |
|---|---|---|---|---|---|---|---|---|---|---|---|---|
| | | | WP | AP | Lv | LP | MP | HP | C | VC | R | |
| 42 | *Chrysopogon aucheri* (Boiss.) Stapf. | Poaceae | X | X | √ | √ | X | X | X | X | √ | All |
| 43 | *Cirsium arvense* (L.) Scope. | Asteraceae | √ | X | X | X | X | √ | √ | X | X | All |
| 44 | *Clematis montana* Buch. | Rananculaceae | X | X | √ | √ | X | X | X | X | √ | Cow, Goat, Sheep |
| 45 | *Clematis napaulensis* Royle | Rananculaceae | X | √ | X | √ | X | X | X | X | √ | Cow, Goat, Sheep |
| 46 | *Colchicum aitchisonii* (Hook. f.) E. Nasir | Liliaceae | X | X | √ | X | X | √ | √ | X | X | Goat |
| 47 | *Colebrookia oppositifolia* Sm. | Lamiaceae | X | X | √ | √ | X | X | √ | X | X | Goat |
| 48 | *Coniogramme rosthornii* Hieron. | Coniogrammaceae | √ | X | X | √ | X | X | √ | X | X | Goat |
| 49 | *Convolvulus arvensis* L. | Convolvulaceae | √ | X | X | X | X | √ | √ | X | X | All |
| 50 | *Conyza aegyptica* Ait. | Asteraceae | X | X | √ | X | √ | X | √ | X | X | Goat, Sheep, Cow |
| 51 | *Conyza bonariensis* L. | Asteraceae | X | X | √ | X | √ | X | √ | X | X | Goat, Sheep, Cow |
| 52 | *Conyza canadensis* L. | Asteraceae | X | X | √ | X | √ | X | √ | X | X | Goat, Sheep, Cow |
| 53 | *Coronopus didymus* (L.) Sm. | Brassicaceae | √ | X | X | X | √ | X | √ | X | X | Goat, Sheep, Cow |
| 54 | *Crotolaria medicagnea* Lam. | Fabaceae | X | √ | X | X | X | √ | √ | X | X | All |
| 55 | *Cuscuta reflexa* Roxb. | Cuscutaceae | √ | X | X | X | X | √ | √ | X | X | Goat, Sheep, Cow |

## Table 7. (Continued)

| S. No. | Plant Name | Family | Parts used | | | Palatability | | | Occurence | | | Animal preference |
|---|---|---|---|---|---|---|---|---|---|---|---|---|
| | | | WP | AP | Lv | LP | MP | HP | C | VC | R | |
| 56 | *Cyperus rotundus* L. | Cyperaceae | X | √ | X | X | X | √ | √ | X | X | All |
| 57 | *Cynodon dactylon* L. Pers | Poaceae | √ | X | X | X | X | √ | X | √ | X | All |
| 58 | *Dactyloctenium aegyptium* L. | Poaceae | X | X | √ | X | X | √ | X | X | √ | All |
| 59 | *Datura stramonium* L. | Solanaceae | X | X | √ | √ | X | X | X | √ | X | Goat |
| 60 | *Debregeasia salicifolia* (D. Don) Rendle | Rhamnaceae | X | √ | X | √ | X | X | √ | X | X | Cow, Goat |
| 61 | *Desmostachya bipinnata* (L.) Stapf | Poaceae | X | X | √ | X | √ | X | √ | X | X | All |
| 62 | *Dicanthium annulatum* (Forssk.) Stapf. | Poaceae | X | X | √ | X | X | √ | √ | X | X | All |
| 63 | *Dicliptera roxburghiana* Nees | Acanthaceae | √ | X | X | X | X | √ | X | √ | X | All |
| 64 | *Digera muricata* (L.) Mart. | Aizoaceae | √ | X | X | X | √ | X | X | √ | X | All |
| 65 | *Dioscorea deltoidea* Wall. ex Kunth | Dioscoreaceae | X | X | √ | X | X | √ | √ | X | X | Goat |
| 66 | *Diospyros lotus* L. | Ebenaceae | X | X | √ | X | X | √ | √ | X | X | Goat |
| 67 | *Dodonaea viscosa* (L.) Jacq. | Sapindaceae | X | √ | X | X | √ | X | X | √ | X | Goat |
| 68 | *Echinochloa crus-galli* (L.) P. Beauv. | Poaceae | √ | X | X | X | X | √ | √ | X | X | All |

| S. No. | Plant Name | Family | Parts used | | | Palatability | | | Occurence | | | Animal preference |
|---|---|---|---|---|---|---|---|---|---|---|---|---|
| | | | WP | AP | Lv | LP | MP | HP | C | VC | R | |
| 69 | *Echinops echinatus* Roxb. | Asteraceae | X | √ | X | X | √ | X | √ | X | X | All |
| 70 | *Eclipta prostrata* (L.) L. | Asteraceae | X | X | X | X | X | √ | X | X | √ | All |
| 71 | *Eragrostis ateroviens* (Desf.) Trin. ex Nees | Poaceae | X | X | √ | X | X | √ | √ | X | X | All |
| 72 | *Eragrostis minor* Host. | Poaceae | X | √ | X | X | X | √ | X | √ | X | All |
| 73 | *Eruca sativa* L. | Brassicaceae | X | X | √ | X | √ | X | X | X | √ | All |
| 74 | *Euphorbia clarkeana* Hkf. | Euphorbiaceae | X | √ | X | X | X | √ | √ | X | X | All |
| 75 | *Euphorbia granulata* Forssk. | Euphorbiaceae | X | √ | X | X | X | √ | √ | X | X | All |
| 76 | *Euphorbia helioscopia* Mewski. | Euphorbiaceae | X | X | √ | X | X | √ | X | √ | X | All |
| 77 | *Euphorbia hirta* L. | Euphorbiaceae | X | X | √ | X | X | √ | √ | X | X | All |
| 78 | *Euphorbia indica* Lam. | Euphorbiaceae | X | X | √ | X | X | √ | √ | X | X | All |
| 79 | *Euphorbia prostrata* (L.) Ait. | Euphorbiaceae | √ | X | X | X | X | √ | √ | X | X | All |
| 80 | *Euphrasia himalayica* Wettst. | Scrophula-riaceae | X | √ | X | X | √ | X | X | √ | X | All |
| 81 | *Ficus carica* L. | Moraceae | X | X | √ | X | √ | X | X | √ | X | Goat |
| 82 | *Ficus bengalensis* L. | Moraceae | X | X | √ | X | √ | X | X | √ | X | Goat |
| 83 | *Ficus palmata* Forssk. | Moraceae | X | X | √ | X | √ | X | X | X | √ | Goat |
| 84 | *Ficus religiosa* L. | Moraceae | X | X | √ | X | √ | X | X | X | √ | Goat, Cow, Donkey |

## Table 7. (Continued)

| S. No. | Plant Name | Family | Parts used | | | Palatability | | | Occurence | | | Animal preference |
|---|---|---|---|---|---|---|---|---|---|---|---|---|
| | | | WP | AP | Lv | LP | MP | HP | C | VC | R | |
| 85 | *Ficus roxburghii* Wall. ex Brand. | Moraceae | X | X | √ | X | √ | X | X | X | √ | Goat |
| 86 | *Flacourtia indica* (Burm. f.) Merr. | Flacourtiaceae | X | X | √ | X | X | √ | √ | X | X | Goat |
| 87 | *Foeniculum vulgare* (Miller) | Apiaceae | √ | X | X | X | X | √ | √ | X | X | All |
| 88 | *Fumaria indica* (Hausskn.) H.N. Pugsley | Fumariaceae | √ | X | X | X | X | √ | X | √ | X | All |
| 89 | *Gallium aparine* L. | Rubiaceae | X | √ | X | X | √ | X | √ | X | X | All |
| 90 | *Geranium rotundifolium* L. | Geraniaceae | X | √ | X | X | X | √ | √ | X | X | Cow, Goat, Sheep |
| 91 | *Grewia optiva* Drum. ex. Burret. | Tiliaceae | X | √ | X | X | X | √ | X | √ | X | All |
| 92 | *Grewia tenax*(Forssk.) Aschers & Schweinf. | Tiliaceae | X | √ | X | X | X | √ | X | √ | X | All |
| 93 | *Hedera nepallensis* K. Koch. | Araliaceae | X | √ | X | X | X | √ | X | √ | X | All |
| 94 | *Heliotopium europaeum* L. | Boraginaceae | X | X | √ | √ | X | X | X | X | √ | Goat, Sheep |
| 95 | *Imperata cylindrica* (L.) Raeuschel | Poaceae | √ | X | X | X | √ | X | √ | X | X | All |
| 96 | *Indigofera himalayensis* Ali | Fabaceae | X | X | √ | X | X | √ | √ | X | X | All |

| S. No. | Plant Name | Family | Parts used | | | Palatability | | | Occurence | | | Animal preference |
|---|---|---|---|---|---|---|---|---|---|---|---|---|
| | | | WP | AP | Lv | LP | MP | HP | C | VC | R | |
| 97 | *Indigofera linifolia* (L. f) Retz. | Fabaceae | √ | X | X | X | X | √ | √ | X | X | All |
| 98 | *Indigofera sessiliflora* DC. | Fabaceae | √ | X | X | X | X | √ | √ | X | X | All |
| 99 | *Ipomoea hederacea* (L.) Jacq. | Convolvu-laceae | √ | X | X | X | X | √ | √ | X | X | All |
| 100 | *Ipomoea nil* (L.) Roth | Convolvu-laceae | √ | X | X | X | X | √ | √ | X | X | All |
| 101 | *Kickxia ramosissima* (Wall.) Janchen | Scrophula-riaceae | X | X | √ | X | X | √ | X | X | √ | Goat, Sheep |
| 102 | *Lactuca auriculata* Wall. ex DC. | Asteraceae | X | X | X | X | X | √ | √ | X | X | All |
| 103 | *Lactuca dissecta* D. Don. | Asteraceae | X | X | √ | X | X | √ | √ | X | X | All |
| 104 | *Lactuca serriola* L. | Asteraceae | √ | X | X | X | X | √ | √ | X | X | All |
| 105 | *Lannea coromandelica* L. | Asteraceae | √ | X | X | X | X | √ | √ | X | X | All |
| 106 | *Lathyrus aphaca* L. | Fabaceae | √ | X | X | X | X | √ | √ | X | X | All |
| 107 | *Launaea procumbens* (Roxb.) Ram. & Rajgo. | Asteraceae | √ | X | X | X | X | √ | √ | X | X | All |
| 108 | *Lepidium sativum* L. | Brassicaceae | √ | X | X | X | X | √ | √ | X | X | All |
| 109 | *Lotus corniculatus* (Wald. & Kit. ex Willd.) Briq. & Rech. F. | Fabaceae | X | √ | X | X | X | √ | X | √ | X | All |
| 110 | *Mallotus philipensis* (Lam.) Muell. | Euphorbiaceae | X | X | √ | X | √ | X | X | √ | X | Goat, Sheep, Cow |

**Table 7. (Continued)**

| S. No. | Plant Name | Family | Parts used | | | Palatability | | | Occurence | | | Animal preference |
|---|---|---|---|---|---|---|---|---|---|---|---|---|
| | | | WP | AP | Lv | LP | MP | HP | C | VC | R | |
| 111 | *Malvastrum coromendelianum* L. | Malvaceae | √ | X | X | X | √ | X | √ | X | X | All |
| 112 | *Maytenus royleanus* (Wall. ex Lawson) Cufodontis | Celastraceae | √ | X | X | X | X | √ | √ | X | X | Goat, Sheep |
| 113 | *Medicago denticulata* Willd. | Fabaceae | √ | X | X | X | X | √ | X | √ | X | All |
| 114 | *Medicago laciniata* (L.) Mill. | Fabaceae | √ | X | X | X | X | √ | X | √ | X | All |
| 115 | *Medicago polymorpha* L. | Fabaceae | √ | X | X | X | X | √ | X | √ | X | All |
| 116 | *Melilotus indica* (L.) All. | Fabaceae | X | X | X | X | X | X | √ | X | X | All |
| 117 | *Mentha longifolia* (L.) Huds | Lamiaceae | X | X | √ | X | X | √ | √ | X | X | Goat |
| 118 | *Micromeria biflora* (Ham.) Bth. | Lamiaceae | X | √ | X | X | √ | X | X | X | √ | All |
| 119 | *Mimosa rubicaulis* Lam. | Mimosaceae | X | √ | X | X | √ | X | X | √ | X | All |
| 120 | *Morus alba* L. | Moraceae | X | X | √ | X | X | √ | √ | X | X | All |
| 121 | *Morus nigra* L. | Moraceae | X | X | √ | X | X | √ | √ | X | X | All |
| 122 | *Myrsine africana* L. | Myrsinaceae | √ | X | X | X | X | √ | √ | X | X | Goat |
| 123 | *Nyctanthes arbor-tristis* L. | Oleaceae | X | X | √ | X | √ | X | √ | X | X | All |
| 124 | *Olea europaea* L. | Oleaceae | X | X | √ | X | √ | X | √ | X | X | All |
| 125 | *Olea ferruginea* Royle | Oleaceae | X | X | √ | X | √ | X | √ | X | X | All |

| S. No. | Plant Name | Family | Parts used | | | Palatability | | | Occurence | | | Animal preference |
|---|---|---|---|---|---|---|---|---|---|---|---|---|
| | | | WP | AP | Lv | LP | MP | HP | C | VC | R | |
| 126 | *Origanum vaulgare* L. | Lamiaceae | X | X | √ | √ | X | X | √ | X | X | Goat |
| 127 | *Otostegia limbata* (Benth.) Boiss. | Lamiaceae | √ | X | X | X | √ | X | X | X | √ | Goat, Sheep |
| 128 | *Oxalis corniculata* L. | Oxalidaceae | X | X | √ | X | X | √ | X | √ | X | Goat, Sheep |
| 129 | *Parthenium hytserophorus* L. | Asteraceae | X | X | X | √ | X | X | √ | X | X | Goat, Sheep |
| 130 | *Peganum harmala* L. | Zygophyllaceae | X | X | √ | √ | X | X | X | √ | X | Goat, Cow, Donkey |
| 131 | *Phalaris minor* Retz. | Poaceae | X | X | √ | X | X | √ | √ | X | X | All |
| 132 | *Phyla nodiflora* (L.) Greene | Verbenaceae | X | √ | | X | X | √ | √ | X | X | All |
| 133 | *Phyllanthus emblica* L. | Euphorbiaceae | X | √ | X | √ | X | X | X | X | √ | Goat, Sheep |
| 134 | *Physalis divaricata* D.Don. | Solanaceae | √ | X | X | X | X | √ | X | X | √ | Goat, Sheep, Cow |
| 135 | *Physalis minima* L. | Solanaceae | √ | X | X | X | X | √ | √ | X | X | All |
| 136 | *Pinus roxburghii* Sargent | Pinaceae | X | X | √ | √ | X | X | X | √ | X | Goat |
| 137 | *Pistacia chinensis Bunge* | Anacardiaceae | X | X | √ | X | X | √ | √ | X | X | Goat, Sheep, Cow |
| 138 | *Plantago lanceolata* L. | Plantginaceae | √ | X | X | X | X | √ | √ | X | X | All |
| 139 | *Plantago major* L. | Plantginaceae | √ | X | X | X | X | √ | √ | X | X | All |
| 140 | *Plantago ovata* Frossk. | Plantginaceae | √ | X | X | X | X | √ | √ | X | X | All |
| 141 | *Polygonum barbatum* L. | Polygonaceae | X | √ | X | X | √ | X | X | X | √ | Goat, Sheep, Cow |

### Table 7. (Continued)

| S. No. | Plant Name | Family | Parts used | | | Palatability | | | Occurence | | | Animal preference |
|---|---|---|---|---|---|---|---|---|---|---|---|---|
| | | | WP | AP | Lv | LP | MP | HP | C | VC | R | |
| 142 | *Polygonum plebejum* R. Br. | Polygonaceae | √ | X | X | X | X | √ | X | √ | X | All |
| 143 | *Polypogon fugax* Nees ex Steud. | Polygonaceae | √ | X | X | X | X | √ | √ | X | X | All |
| 144 | *Pongamia pinnata*(L.) Pierre | Papilionaceae | X | X | √ | X | √ | X | X | √ | X | All |
| 145 | *Populus deltoides* Bartram ex Marsh. | Salicaceae | X | X | √ | √ | X | X | √ | X | X | Goat, Sheep, Cow |
| 146 | *Potentilla supina* L. | Rosaceae | X | X | √ | X | √ | X | √ | X | X | Goat, Sheep, Cow |
| 147 | *Pteridium aquilinum* (L.) Kuhn | Pteridaceae | √ | X | X | X | X | √ | X | X | √ | Goat, Sheep, Cow |
| 148 | *Psammogeton biternatum* Edgew. | Papilionaceae | √ | X | X | X | X | √ | √ | X | X | All |
| 149 | *Pueraria tuberosa* (Roxb. Ex Willd) DC. | Apiaceae | √ | X | X | X | X | √ | √ | X | X | All |
| 150 | *Punica granatum* L. | Punicaceae | X | X | √ | X | X | √ | X | √ | X | Goat, Sheep, Cow |
| 151 | *Pyrus pashia* Ham. ex D. Don | Rosaceae | X | X | √ | X | X | √ | X | √ | X | Goat, Sheep, Cow |
| 152 | *Quercus dilatata* Lindley | Fagaceae | X | X | √ | √ | X | X | X | X | √ | All |
| 153 | *Quercus incana* Roxb. | Fagaceae | X | X | √ | √ | X | X | X | X | √ | All |
| 154 | *Quercus leucotrichophora* | Fagaceae | X | X | √ | √ | X | X | X | X | √ | All |

| S. No. | Plant Name | Family | Parts used | | | Palatability | | | Occurence | | | Animal preference |
|---|---|---|---|---|---|---|---|---|---|---|---|---|
| | | | WP | AP | Lv | LP | MP | HP | C | VC | R | |
| 155 | *Rananculus sceleratus* L. | Rananculaceae | X | √ | X | √ | X | X | √ | X | X | All |
| 156 | *Ranunculus arvensis* L. | Rananculaceae | X | √ | X | √ | X | X | √ | X | X | All |
| 157 | *Rananculus muricatus* L. | Rananculaceae | X | √ | X | √ | X | X | √ | X | X | All |
| 158 | *Rhynchosia minima* (L.) DC | Fabaceae | √ | X | X | X | X | √ | √ | X | X | All |
| 159 | *Rosa brunonii* Lindll. | Rosaceae | X | √ | X | X | √ | X | X | X | √ | Goat |
| 160 | *Rubia cordifolia* L. | Rubiaceae | X | √ | X | X | √ | X | X | √ | X | Goat |
| 161 | *Rubus ellipticus* Smith | Rubiaceae | √ | X | X | X | √ | X | √ | X | X | Goat |
| 162 | *Rumex dentatus* L. | Polygonaceae | X | X | √ | X | X | √ | X | √ | X | All |
| 163 | *Rumex hastatus* D. Don. | Polygonaceae | X | X | √ | X | X | √ | X | √ | X | Goat |
| 164 | *Rumex nepalensis* Spreng | Polygonaceae | X | X | √ | X | √ | X | √ | X | X | Goat |
| 165 | *Sageretia thea* (Osback) M. C. Johnston | Rhamnaceae | X | X | √ | X | X | √ | X | √ | X | X |
| 166 | *Saccharum bengalense* Retz. | Poaceae | X | X | √ | X | √ | X | X | X | √ | Cow, Donkey |
| 167 | *Saccharum spontaneum* L. | Poaceae | X | X | √ | X | √ | X | √ | X | X | Cow, Donkey |
| 168 | *Salvia moorcroftiana* Wall. ex Bth. | Lamiaceae | √ | X | X | X | X | √ | X | √ | X | All |
| 169 | *Salvia plebeia* R. Br. | Lamiaceae | √ | X | X | X | X | √ | X | √ | X | All |

**Table 7. (Continued)**

| S. No. | Plant Name | Family | Parts used | | | Palatability | | | Occurence | | | Animal preference |
|---|---|---|---|---|---|---|---|---|---|---|---|---|
| | | | WP | AP | Lv | LP | MP | HP | C | VC | R | |
| 170 | *Sauromatum venosum* (Ait.) Schott. | Araceae | X | X | √ | √ | X | X | X | X | √ | Goat |
| 171 | *Saussurea albescens* (DC.) Schr. Bip. | Asteraceae | X | √ | X | X | X | √ | √ | X | X | All |
| 172 | *Saussurea atkinsonii* (Clarke) | Asteraceae | X | √ | X | X | X | √ | √ | X | X | All |
| 173 | *Saussurea heteromalla* DC. | Asteraceae | √ | X | X | X | X | √ | √ | X | X | All |
| 174 | *Setaria glauca* (L.) P.Beauv | Poaceae | X | X | √ | X | X | √ | X | X | √ | All |
| 175 | *Sida cordata* (Burm. f.) Borss.-Waalkes | Poaceae | X | X | √ | X | X | √ | X | X | √ | Goat, Sheep, Cow |
| 176 | *Silene conoidae* L. | Malvaceae | X | √ | X | X | X | √ | √ | X | X | All |
| 177 | *Silybum marianum* (L.) Gaertin | Caryophyllaceae | X | X | √ | √ | X | X | X | X | √ | Goat, Sheep, Cow |
| 178 | *Sisymbrium irio* L. | Asteraceae | X | X | X | X | X | √ | X | √ | X | All |
| 179 | *Smilax aspera* L. | Smilacaceae | X | X | √ | X | X | X | X | X | √ | All |
| 180 | *Solanum incanum* L. | Solanaceae | √ | X | X | X | X | √ | √ | X | X | All |
| 181 | *Solanum nigrum* L. | Solanaceae | √ | X | X | X | X | √ | √ | X | X | All |
| 182 | *Solanum surattense* Burm.f. | Solanaceae | √ | X | X | X | X | √ | √ | X | X | Goat, Sheep |
| 183 | *Solanum villosum* (L.) Moench | Solanaceae | √ | X | X | X | X | √ | X | √ | X | All |
| 184 | *Sonchus asper* (L.) Hill. | Solanaceae | X | X | X | X | X | √ | √ | X | X | All |
| 185 | *Sonchus arvense* L. | Solanaceae | √ | X | X | X | X | √ | X | √ | X | All |

| S. No. | Plant Name | Family | Parts used | | | Palatability | | | Occurence | | | Animal preference |
|---|---|---|---|---|---|---|---|---|---|---|---|---|
| | | | WP | AP | Lv | LP | MP | HP | C | VC | R | |
| 186 | *Sorghum bicolor* (L.) Moench. | Asteraceae | X | X | √ | X | X | √ | √ | X | X | All |
| 187 | *Sorghum halepense* (L.) Bern. | Poaceae | X | X | √ | √ | X | X | X | X | √ | Goat |
| 188 | *Stellaria media* (L.) Cry. | Caryophyllaceae | √ | X | X | X | X | √ | X | √ | X | All |
| 189 | *Taraxacum officinale* Weber. | Asteraceae | X | X | √ | X | X | √ | X | √ | X | All |
| 190 | *Taraxcum wallichii* DC. | Asteraceae | √ | X | X | X | X | √ | √ | X | X | All |
| 191 | *Themeda anathera* (Nees) Hack | Poaceae | √ | X | X | X | X | √ | X | √ | X | All |
| 192 | *Trianthema portulacastrum* L. | Aizoaceae | √ | X | X | X | √ | X | X | X | √ | All |
| 193 | *Tribulus terrestris* L. | Zygophyllaceae | √ | X | X | X | X | √ | √ | X | X | All |
| 194 | *Trichodesma indicum* (L.) R. Br. | Boraginaceae | X | X | √ | √ | X | X | X | X | √ | Goat, Sheep |
| 195 | *Trifolium repens* L. | Fabaceae | √ | X | X | X | X | √ | X | √ | X | Goat, Sheep |
| 196 | *Valeriana wallichii* DC. | Valerianaceae | X | √ | X | X | X | √ | X | √ | X | Goat, Sheep, Cow |
| 197 | *Verbascum thapsus* L. | Scrophulariaceae | √ | X | X | X | X | √ | √ | X | X | Goat, Sheep, Cow |
| 198 | *Verbena officinalis* L. | Verbenaceae | √ | X | X | X | X | √ | X | √ | X | All |
| 199 | *Vicia faba* L. | Fabaceae | √ | X | X | X | X | √ | √ | X | X | All |
| 200 | *Vicia sativa* L. | Fabaceae | √ | X | X | X | X | √ | √ | X | X | All |
| 201 | *Viola cancescens* Wall. ex Roxb. | Violaceae | √ | X | X | X | X | √ | √ | X | X | All |

**Table 7. (Continued)**

| S. No. | Plant Name | Family | Parts used | | | Palatability | | | Occurence | | | Animal preference |
|---|---|---|---|---|---|---|---|---|---|---|---|---|
| | | | WP | AP | Lv | LP | MP | HP | C | VC | R | |
| 202 | *Vitex negundo* L. | Verbenaceae | X | √ | X | X | √ | X | √ | X | X | Goat, Sheep, Cow |
| 203 | *Withania somnifera* (L.) Dunal. | Simarubaceae | √ | X | X | X | X | √ | √ | X | X | Goat |
| 204 | *Woodfordia fruticosa* (L.) S. Kurz | Vitaceae | X | X | √ | X | X | √ | X | √ | X | All |
| 205 | *Zanthoxylum alatum* Roxb. | Rutaceae | X | X | √ | X | √ | X | X | √ | X | Goat, Sheep |
| 206 | *Zizyphus mauritiana* Mill. | Rhamnaceae | X | X | √ | X | X | √ | √ | X | X | All |
| 207 | *Zizyphus oxyphylla* Edgew. | Rhamnaceae | X | X | √ | √ | X | X | X | X | √ | All |
| | | | 61 | 33 | 68 | 26 | 37 | 106 | 97 | 34 | 38 | |

## 4.1.12. Mode of Application

Ten different forms of recipes were employed by the inhabitants of area for treating various diseases. The most preferred recipe form is decoction (29 spp., 22.14%), followed by powder (28 spp., 21.37%), raw form (22 spp., 16.79%), juice (16 spp., 12.21%), extract (13 spp., 9.92%) and paste (11 spp., 8.40%), whereas, four other forms are less used (Table 6).

# 4.2. FODDER/FORAGE

A total of 207 species are used for multiple purposes by the inhabitants of the study area. The inventory is alphabetically arranged and ptovided in Table 7.

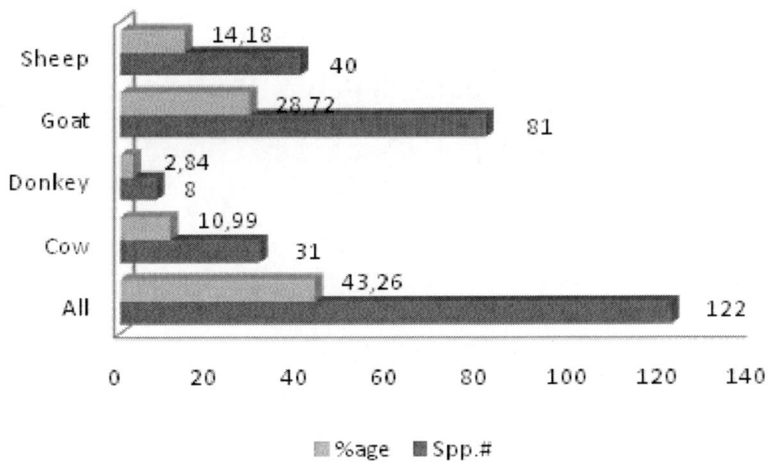


Figure 2. Animal preference for feeding different plants of the study area.

## 4.2.1. Animal Preference

The study area is rangeland and cannot be converted into agricultural land; therefore people normally domesticate four animals viz., Goat, Sheep, Cows and Donkey for earning their livelihood. The area is endowed with high proportion of palatable species. There was highest number of species (122 species, 43.26%) were found palatable for all those animals (Figure 2). Besides, Goat was the most preferred and suited animal to the climatic conditions as browser which preferred 81 species (28.72%). Sheep was found attached with 40 species (14.18%), whereas, Cows alone utilized 31 species (10.99%).

### 4.2.2. Plant Parts

All animals prefer leaves (42%) might be due to presence of ground flora and trees and shrubs (Figure 3). It was followed by whole plant (38%) and aerial parts (20%).

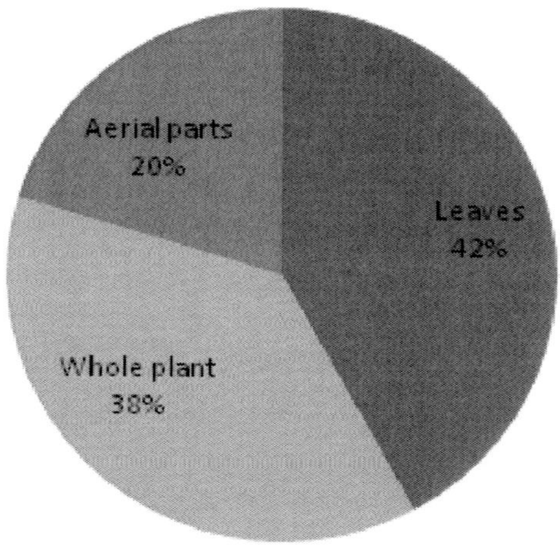


Figure 3. Different parts of the plants grazed/browsed by various animals.

### 4.2.3. Forage Availability Round the Year

The area sustains good diversity of palatable species available almost round the year. Most of flora bloom in the month of April and about 136 plant species (60.44%) were available for the livestock (Figure 4). The second important month was March in which 126 species (56%) were found in flowering condition, flowed by May (117 spp., 52%), August (101 spp., 44.89%), September (99 spp., 44%), June (96 spp., 42.67%), October (92 spp., 40.89%) and July (90 spp., 40%), whereas rest of four months viz., Febuary,

November, December and January within the range of 65-83 species. The palatable flora was gradullay decline after month of July.

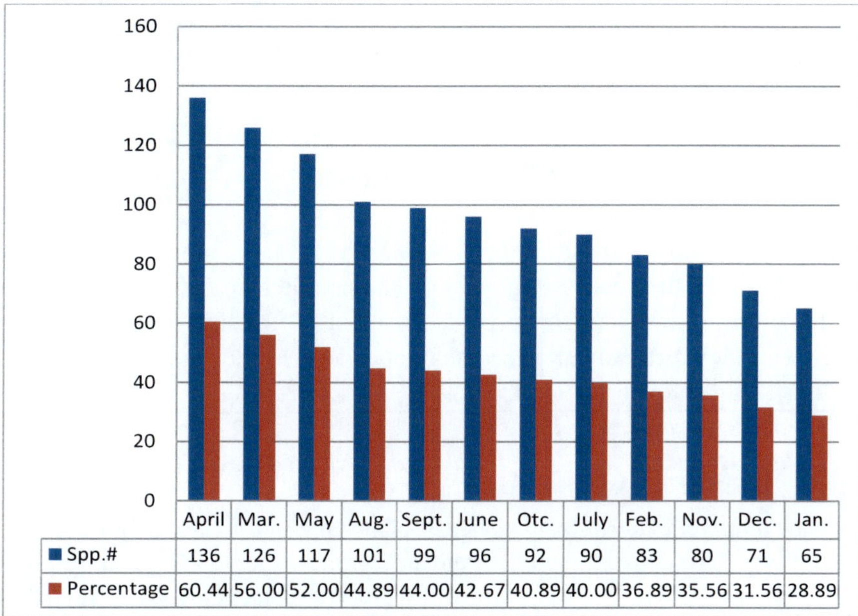


| | April | Mar. | May | Aug. | Sept. | June | Otc. | July | Feb. | Nov. | Dec. | Jan. |
|---|---|---|---|---|---|---|---|---|---|---|---|---|
| Spp.# | 136 | 126 | 117 | 101 | 99 | 96 | 92 | 90 | 83 | 80 | 71 | 65 |
| Percentage | 60.44 | 56.00 | 52.00 | 44.89 | 44.00 | 42.67 | 40.89 | 40.00 | 36.89 | 35.56 | 31.56 | 28.89 |

Figure 4. Availability of forage during different months of the year.

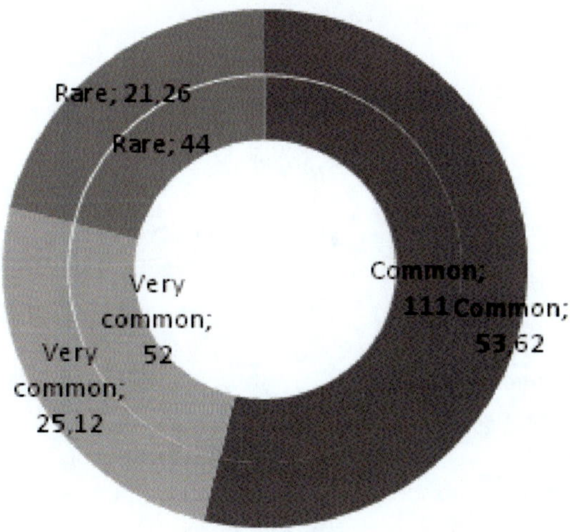


Figure 5. Status of forage species in the study area.

### 4.2.4. Occurrence of Forage Species

There was highest proportion of commonly growing species in the area (111 spp., 58.62%), followed by very common (52 spp., 25.12%), whereas, 44 species (21.26%) were marked as rare (Figure 5).

## 4.3. FUEL WOOD

This is second most important need of the human being after fodder/forage for the energy source, since the people faces severe cold season. For this purpose, they collect wood from about 67 species. The inventory is family-wise alphabetically arranged and appended as follows:

| | | |
|---|---|---|
| | *Family: Acanthaceae* | |
| 1 | *Botanical name:* | *Adhatoda zeylanica* Medic. |
| | *Vernacular name:* | Baikar |
| | *Other multifarious use:* | This plant present commonly in the Potohar region. On foot hills and hilly site. Flower gives sweetness which is used in sweet dishes. Unripe bananas are rolled in the leaves of this plant for ripening. Plant is used as fuel. |
| | *Family: Anacardiaceae* | |
| 2 | *Botanical name:* | *Pistacia chinensis* Bunge |
| | *Vernacular name:* | Kangar |
| | *Other multifarious use:* | It is commonly present on hilly and cold areas on the rocky top mountains. Wood of the tree is used for fuel purpose. |
| | *Family: Apocynaceae* | |
| 3 | *Botanical name:* | *Carissa opaca* Stapf ex. Haines |
| | *Vernacular name:* | Granda |
| | *Part used:* | Leaves, branches, root, fruit |
| | *Other multifarious use:* | This is small tree and its branches are used for fuel purpose. |
| 4 | *Botanical name:* | *Nerium oleander* L. |
| | *Vernacular name:* | Kanair |

| | | |
|---|---|---|
| | *Other multifarious use:* | It is shrubby plant present mostly along the road side. Plant has showy flowers that are why Plant is growing as an ornamental. |
| | *Family: Asclepiadaceae* | |
| 5 | *Botanical name:* | *Calotropis procera* (Willd.) R. Br. |
| | *Vernacular name:* | Aak, Madar |
| | *Other multifarious use:* | This is commonly found species in study area. It is also present in desert habitat. Plant stems are used as fuel. |
| 6 | *Botanical name:* | *Periploca aphylla* Decne. |
| | *Vernacular name:* | Baata |
| | *Other multifarious use:* | This is fibrous shrub, Stem fibers are used for making ropes, thatching huts. Flowers buds are rarely eaten as raw. Plant stems are used as fuel. |
| 7 | *Botanical name:* | *Carthamus oxycantha* M. Bieb. |
| | *Vernacular name:* | Pholi |
| | *Other multifarious use:* | It is very common spiny herb mostly present in the month of April to May. The plant in juvenile stage is dried form fuel use. |
| 8 | *Botanical name:* | *Tagetes minuta* L. |
| | *Vernacular name:* | Sadbarga |
| | *Other multifarious use:* | The plant is also grown as ornamental having yellow showy flowers. The plant is used as fuel source |
| 9 | *Botanical name:* | *Xanthium strumarium* L. |
| | *Vernacular name:* | Bhurt |
| | *Other multifarious use:* | It is spiny shrub mostly present on the foot hill of the study area. The plant is used as fuel source. |
| | *Family:Berberidaceae* | |
| 10 | *Botanical name:* | *Berberis lycium* Royle. |
| | *Vernacular name:* | Sumbulu |
| | *Other multifarious use:* | It is spiny yellow flower herbaceous plant. Leaves are used for making Chatni. The plant is sued as fuel source. |
| | *Family: Brassicaceae* | |
| 11 | *Botanical name:* | *Eruca sativa* L. |
| | *Vernacular name:* | Tara Meera |

**(Continued)**

| | | |
|---|---|---|
| | *Other multifarious use:* | It is herbaceous oily seeds plant. The dried plant is used as fuel wood source |
| | *Family: Bombacaceae* | |
| 12 | *Botanical name:* | *Bombax ceiba* L. |
| | *Vernacular name:* | Simbal |
| | *Other multifarious use:* | The wood is used as fuel. The silky floss of seeds is used for stiffing pillows and quills. It is every green tree. The silky floss of seeds is used for stiffing pillows and quills. The plant is used as fuel. |
| 13 | *Botanical name:* | *Maytenus royleanus* (Wall. ex Lawson) Cufodontis |
| | *Vernacular name:* | Patakhi |
| | *Other multifarious use:* | It is ever green spiny shrub. The branches are used to make walking sticks. Plant is used as fuel wood source. |
| 13 | *Botanical name:* | *Dioscorea deltoides* Wall. ex Kunth |
| | *Vernacular name:* | ----- |
| | *Other multifarious use:* | This plant grows in Sandy loam to humid rich soil. Branches are used as fuel source. |
| | *Family: Ebenaceae* | |
| 14 | *Botanical name:* | *Diospyros lotus* L. |
| | *Vernacular name:* | Amlok |
| | *Other multifarious use:* | It is rare plant present on mountain top. It has wild fruit which are highly nutritious. Theses Fruits are eaten as raw due to sour taste The dried branches are used fuel. |
| | *Family: Euphorbiaceae* | |
| 15 | *Botanical name:* | *Mallotus philipensis* (Lam.) Muell. |
| | *Vernacular name:* | Kamela |
| | *Other multifarious use:* | It is tall woody tree along the road side of the study area. Its wood is supposed having strong fibre and used for making furniture. Trunk is used as fuel source. |
| 16 | *Botanical name:* | *Phyllanthus emblica* L. |
| | *Vernacular name:* | Amla |

| | | |
|---|---|---|
| | *Other multifarious use:* | It is highly medicinal plant rarely present in the study area. Wood is used as fuel for furniture making and construction. Leaves soaked in water along with soapnut and Shika kai and applied on hairs for long and healty hairs. |
| 17 | *Botanical name:* | *Ricinus communis* L. |
| | *Vernacular name:* | Arind/Harnoli |
| | *Other multifarious use:* | This is small tree poisonus plant. The wood is as fuel purpose. |
| | *Family: Fabaceae* | |
| 18 | *Botanical name:* | *Indigofera himalayensis* Ali |
| | *Vernacular name:* | ----- |
| | *Other multifarious use:* | It is shrubby plant mostly present in deserted and sandy areas during July to October. The plant is used as fuel purpose. |
| | *Family: Fagaceae* | |
| 19 | *Botanical name:* | *Quercus dilatata* Lindl. |
| | *Vernacular name:* | Barungi |
| | *Other multifarious use:* | It is evergreen tree commonly present all over the mountainous side of the study area its flowering season is April to May. Wood is used as furniture and fuel purposes |
| 20 | *Botanical name:* | *Quercus incana* Roxb. |
| | *Vernacular name:* | Rein, Shah-e-baloot |
| | *Other multifarious use:* | It is evergreen tree commonly present all over the mountainous side of the study area. Wood is used as furniture and fuel purposes |
| 21 | *Botanical name:* | *Quercus leucotrichophora* A. Camus |
| | *Vernacular name:* | Rein, Shah-e-baloot |
| | *Other multifarious use:* | It is evergreen tree commonly present all over the mountainous side of the study area. The wood is used for construction and fuel purposes |
| 22 | *Family: Flacourtiaceae* | |
| | *Botanical name:* | *Flacourtia indica* (Burm. f.) Merrill |
| | *Vernacular name:* | Kokoh |
| | *Other multifarious use:* | It is highly medicinal tree which have wild fruits theses fruits are edible. The wood is used in building material. The plant is used as fuel after drying. |

**(Continued)**

| | | |
|---|---|---|
| | *Family: Lamiaceae* | |
| 23 | *Botanical name:* | *Colebrookia oppositifolia* Sm. |
| | *Vernacular name:* | Shakar Dana |
| | *Other multifarious use:* | This small shrub present dominantly on hilly site. The branches are used as fuel. |
| | *Family: Malvaceae* | |
| 24 | *Botanical name:* | *Abutilon bidentatum* Hochst. ex Rich. |
| | *Vernacular name:* | Kanghi buti |
| | *Other multifarious use:* | This plant usually presents in April to October. It have showy yellow flower. The branches are used as fuel. |
| | *Family: Meliacea* | |
| 25 | *Botanical name:* | *Melia azedarach* L. |
| | *Vernacular name:* | Dharek |
| | *Other multifarious use:* | It is highly medicinal plant present in the Potohar Region commonly. Leaves of plant used in rice field (*Paneeri*) acting as antifeedant. The same is also applied in storage place of wheat and rice to prevent from insect. |
| | *Family: Mimosaceae* | |
| 26 | *Botanical name:* | *Acacia modesta* Wall. |
| | *Vernacular name:* | Phulai |
| | *Other multifarious use:* | It is evergreen spiny tree which is mostly present on sandy and temperate region. Plant wood is very strong and used for construction proposes and furniture material. Wood is used as fuel. |
| 27 | *Botanical name:* | *Albizzia lebbeck* (L.) Benth. |
| | *Vernacular name:* | Shirin |
| | *Other multifarious use:* | It is tall wood tree, inflorescence appear in the month of March to May. Wood of the plant is very strong is used for construction and furniture purpose. The wood is also used for fuel purpose. |
| 28 | *Botanical name:* | *Mimosa rubicaulis* Lam. |
| | *Vernacular name:* | Ral |
| | *Other multifarious use:* | It is highly medicinal plant rarely present in the study area. Its branches used as fuel. |

| | | |
|---|---|---|
| | *Family: Moraceae* | |
| 29 | *Botanical name:* | *Ficus bengalensis* L. |
| | *Vernacular name:* | Barr |
| | *Other multifarious use:* | It is commonly present plant in the study area having wild fruits. These fruits eaten as a raw. Branches used as fuel wood source. |
| 30 | *Botanical name:* | *Ficus carica* L. |
| | *Vernacular name:* | Anjeer/Phuwari/Phagwari |
| | *Other multifarious use:* | It is commonly present plant in the study area having wild fruits. It fruits are edible. Young leaves are cooked as vegetable. Wood is used as fuel and roof thatching. |
| 31 | *Botanical name:* | *Ficus palmata* Forssk. |
| | *Vernacular name:* | Phagwara |
| | *Other multifarious use:* | It is commonly present plant in the study area having wild fruits. It fruits are edible. Young leaves are cooked as vegetable. Wood is used as fuel and roof thatching and building material. |
| 32 | *Botanical name:* | *Ficus roxburghii* Wall. ex Brand. |
| | *Vernacular name:* | Dusi |
| | *Other multifarious use:* | It is commonly present plant in the study area having wild fruits. It fruits are edible. Young leaves are cooked as vegetable. Wood is used as fuel and roof thatching and building material. |
| 33 | *Botanical name:* | *Morus alba* L. |
| | *Vernacular name:* | Shehtoot |
| | *Other multifarious use:* | This is highly medicinal and useful plant. Fruits are usually appearing in the month of April to September which are edible. Wood is durable and used for furniture making and as fuel purpose. |
| 34 | *Botanical name:* | *Morus nigra* L. |
| | *Vernacular name:* | Tut |
| | *Other multifarious use:* | This is highly medicinal and useful plant. Fruits are usually appearing in the month of April to September which are edible is very delicious and eaten raw. Wood is durable and used for furniture making and as fuel purpose. Wood is used for construction propose. |

**(Continued)**

| | | |
|---|---|---|
| | *Family: Myrsinaceae* | |
| 35 | *Botanical name:* | *Myrsine africana* L. |
| | *Vernacular name:* | Khokhal/Khokhan |
| | *Other multifarious use:* | This plant bearing wild fruits, these fruits are edible, very delicious and eaten raw. The dried plant is used as fuel. |
| | *Family: Oleaceae* | |
| 36 | *Botanical name:* | *Nyctanthes arbor-tristis* L. |
| | *Vernacular name:* | Harsingar |
| | *Other multifarious use:* | It is shrubby plant usually appear in the month of August to October. The plant possesses aromatic and fragrant flowers. |
| 37 | *Botanical name:* | *Olea europaea* L. |
| | *Vernacular name:* | Koh |
| | *Other multifarious use:* | It is tall tree plant stems are used as fuel wood. The wood of this plant is very strong and used for construction proposes like roof thatching and furniture making. |
| 38 | *Botanical Name:* | *Olea ferruginea* L. |
| | *Vernacular name:* | Kahu |
| | *Other multifarious use:* | It is tall tree plant stems are used as fuel wood. The wood of this plant is very strong and used for construction proposes like roof thatching and furniture making. Dried leaves are used for tea making |
| | *Family: Papilionaceae* | |
| 39 | *Botanical name:* | *Pongamia pinnata* (L.) Pierre |
| | *Vernacular name:* | Sukh chain |
| | *Other multifarious use:* | It is found growing along the road site of the study and also present on the hilly site. The wood is used in Furniture and fuel. |
| | *Family: Pinaceae* | |
| 40 | *Botanical name:* | *Pinus roxburghii* Sargent |
| | *Vernacular name:* | Chir |

| | | |
|---|---|---|
| | *Other multifarious use:* | This is gymnosperm evergreen dominant tree. Having Resin in the plant. Resin (ganda beroza) is Used in paintings. Seeds are edible. Dry leaves are used for thatching of roofs. The braches and wood is used for fuel source. Leaves supposed to have allelopathic activities. Due to resin its wood is highly flammable. |
| | *Family: Poaceae* | |
| 41 | *Botanical name:* | *Desmostachya bipinnata* (L.) *S*tapf. |
| | *Vernacular name:* | Dab Ghaa |
| | *Other multifarious use:* | This is grass that culm is used as broom making. Leaves are woven to make ropes. It is used as fuel proposes after drying. |
| 42 | *Botanical name:* | *Dicanthium annulatum* (Forssk.) Stapf. |
| | *Vernacular name:* | Murgha Ghaa |
| | *Other multifarious use:* | This is very common grass in the study area whose flowering periods is March to April. It is also used as fuel proposes after drying. |
| 43 | *Botanical name:* | *Saccharum bengalense* Retz. |
| | *Vernacular name:* | Kana |
| | *Other multifarious use:* | This is commonly found grass in the study areas. The leaves are woven to make matrices. The stem/culm is used for making matrices, sitting chairs, etc. (Patlan, chick, Moorhe) and other decoration pieces (e.g. cage, boxes, etc). Inflorescence is used for making brooms (*Jharoo*). Plant stem is used for making pencils (*Qalam*) and supporting materials. Leaves are grazed/browsed by cattle. The plant is used as fuel source. The culm is used for thatching of roofs as well as making *Chiks*, chairs (*Moorhe*), and decoration. |
| 44 | *Botanical name:* | *Saccharum spontaneum* L. |
| | *Vernacular name:* | Kanna |

**(Continued)**

| | | |
|---|---|---|
| | *Other multifarious use:* | This is commonly found grass in the study areas. The leaves are woven to make matrices. The stem/culm is used for making matrices, sitting chairs, etc. (Patlan, chick, Moorhe) and other decoration pieces (e.g. cage, boxes, etc). Inflorescence is used for making brooms (*Jharoo*). Plant stem is used for making pencils (*Qalam*) and supporting materials. Leaves are grazed/browsed by cattle. The plant is used as fuel source. The culm is used for thatching of roofs as well as making *Chiks*, chairs (*Moorhe*), and decoration. |
| 45 | *Botanical name:* | *Themeda anathera* (Nees) Hack. |
| | *Vernacular name:* | Loonder, Lunji |
| | *Other multifarious use:* | This is commonly found grass in the hilly site of the study areas. Leaves are used for thatching of roof. Plant in fresh form is used as favorite fodder for cattle. The same is also collected and stored for autumn/winter season. The plant is good soil binder. Culms are used in broom making. |
| | *Family: Punicaceae* | |
| 46 | *Botanical name:* | *Punica granatum* L. |
| | *Vernacular name:* | Druna/Druni |
| | *Other multifarious use:* | This is highly medicinal plant commonly present in the study area. Seeds are used as flavoring agent in cooking food material. Plant is wood is used as fuel material. |
| | *Family: Rhamnaceae* | |
| 47 | *Botanical name:* | *Sageretia thea* (Osback) M. C. Johnston |
| | *Vernacular name:* | Chir Nakhtar |
| | *Other multifarious use:* | This rare plant in the study area having raw fruit. Fruit is edible. Wood of the plant is use for fuel purpose. |
| 48 | *Botanical name:* | *Zizyphus mauritiana* Mill. |
| | *Vernacular name:* | Beri |
| | *Other multifarious use:* | This small spiny tree plant having wild fruits, these Fruits known as *Ber* which are edible. Its wood is very strong and used in furniture and in construction. Wood is used as fuel source. |

| | | |
|---|---|---|
| 49 | *Botanical name:* | *Zizyphus oxyphylla* Edgew. |
| | *Vernacular name:* | Ber maloki |
| | *Other multifarious use:* | This small spiny tree plant having wild fruits, these Fruits known as *Ber* which are edible. Its wood is very strong and used in furniture and in construction. Wood is used as fuel source. |
| | *Family: Rosaceae* | |
| 50 | *Botanical name:* | *Rosa brunonii* Lindll. |
| | *Vernacular name:* | Jangli gulab |
| | *Other multifarious use:* | This is spiny plant; This plant is growing for the beautification around settlements. Plants are used as fodder and fuel. It is also growing as hedge. |
| 51 | *Botanical name:* | *Pyrus pashia* Ham. ex D. Don |
| | *Vernacular name:* | Batangi |
| | *Other multifarious use:* | This is wild fruiting plant and found rarely present in the study area. The wood is used for making sticks and as fuel. Ripened fruit is eaten as raw. |
| 52 | *Botanical name:* | *Rubus ellipticus* Smith |
| | *Vernacular name:* | Aakhra |
| | *Other multifarious use:* | This is wild fruiting plant and found rarely present in the study area. Its fruit which is called *Aakha* used as delicious fruit. Plant used as fuel source. |
| | *Family: Rutaceae* | |
| 53 | *Botanical name:* | *Zanthoxylum alatum* Roxb |
| | *Vernacular name:* | Timbar/Timar |
| | *Other multifarious use:* | This is highly medicinal plant it is commonly present in the study area. Seeds are used as substitute of black pepper (*Piper nigrum*) and for making *Chatni*. This plant is used as fuel wood species. |
| | *Family: Salicaceae* | |
| 54 | *Botanical name:* | *Populus deltoides* Bartram ex Marsh. |
| | *Vernacular name:* | Sufaid poplar |
| | *Other multifarious use:* | This is tall evergreen plant its branches are used for making agricultural tools and construction (Shattering) material. Wood is used as fuel source. |
| | *Family: Sapindaceae* | |

**(Continued)**

| | | |
|---|---|---|
| 55 | *Botanical name:* | *Dodonaea viscosa* (L.) Jacq. |
| | *Vernacular name:* | Snatha |
| | *Other multifarious use:* | This is evergreen bushy plant. Plant mostly used as hedge bush. The branches are used in thatching of roof. The dried plant is used as fuel. |
| | *Family: Scrophulariaceae* | |
| 56 | *Botanical name:* | *Verbascum thapsus* L. |
| | *Vernacular name:* | Pahari Tambaku |
| | *Other multifarious use:* | This is highly medicinal plant usually found in the month of March to October. Leaves are used as substitute of cigarette (*Hukka*). Plant is used fuel source. |
| | *Family: Simarubaceae* | |
| 57 | *Botanical name:* | *Ailanthus altissima* (Mill.) Swingle |
| | *Vernacular name:* | Durawia |
| | *Other multifarious use:* | This is dominant plant in different patches of the study area. It is evergreen tall tree. The straight branches are used in shattering of building. The plant is a good source of fuel wood. |
| | *Family: Solanaceae* | |
| 58 | *Botanical name:* | *Datura stramonium* L. |
| | *Vernacular name:* | Dhatura |
| | *Other multifarious use:* | These plants contain milky latex. It is commonly present shrub in the Potohar Region. The plant is acting as shade plant and fuel wood source. |
| 59 | *Botanical name:* | *Withania somnifera* (L.) Dunal. |
| | *Vernacular name:* | Asghand/Aksan |
| | *Other multifarious use:* | This is highly medicinal shrubby plant commonly present in the study area. Plant is used for fuel wood. |
| | *Family: Tiliaceae* | |
| 60 | *Botanical name:* | *Grewia optiva* Drum. ex Burret. |
| | *Vernacular name:* | Taman |
| | *Other multifarious use:* | This rarely present evergreen tall tree. This is very important plant in ethnobotanical and medicinal uses references. Twigs are soaked in water for some days. These are then beaten to remove bark which is used in making ropes. Wood/twigs are used for fuel wood source. |

| | | |
|---|---|---|
| 61 | *Botanical name:* | *Grewia tenax* (Forssk.) Aschers & Schweinf. |
| | *Vernacular name:* | Gunger |
| | *Other multifarious use:* | This is bushy shrub rarely present in the present in the study area. Stems are used as fuel source. |
| | *Family: Umbelliferae* | |
| 62 | *Botanical name:* | *Foeniculum vulgare* Miller |
| | *Vernacular name:* | Soonuf |
| | *Other multifarious use:* | This Plant is highly medicinal and very important in medicinal references. Plant extract is used in as flavouring agent, spices and different beverages. The branches used as fuel wood. |
| | *Family: Urticaceae* | |
| 63 | *Botanical name:* | *Debregeasia salicifolia* (D. Don) Rendle |
| | *Vernacular name:* | Sinduri/Sindur |
| | *Other multifarious use:* | This is evergreen tree wood species. Its flowering season is mostly February to April. The dried plant is used as fuel wood. Branches are used in roof thatching because its wood is insect resistant. |
| | *Family: Verbenaceae* | |
| 64 | *Botanical name:* | *Lantana camara* L. |
| | *Vernacular name:* | Panch phul |
| | *Other multifarious use:* | This invasive evergreen species. It has showy flowers, mostly grow for ornamental purpose. The plant is used as fuel wood. The same is supposed to be poisonous for cattle. |
| 65 | *Botanical name:* | *Lantana indica* Roxb. |
| | *Vernacular name:* | Ghaneri |
| | *Other multifarious use:* | This evergreen shrub species which is rarely present in the study area. The dried branches are used as fuel wood. |
| 66 | *Botanical name:* | *Vitex negundo* L. |
| | *Vernacular name:* | Banna |
| | *Other multifarious use:* | This is evergreen small tree. The plant is grown along water channels to check erosion. Leaves are used for grains storage supposed to possess insecticidal properties. The young shoots are used basket making as well as tooth brush for cleaning teeth. Branches of plant are used for fuel purpose. |
| | *Family: Violaceae* | |
| 67 | *Botanical name:* | *Woodfordia fruticosa* (L.) S. Kurz |
| | *Vernacular name:* | Tavi |
| | *Other multifarious use:* | It is tall tree evergreen species. The plant is used as fuel wood source. |

## 4.4. Multifarious Use

A total of 75 species are used for multiple purposes by the inhabitants of the study area. The inventory is alphabetically arranged and appended as follows:

## 4.5. Ethnotaxonomic Approach

The naming of plants is an old tradition exists in almost all indigenous communities of the world reflecting their dependence on them since the earliest stages of civilization. As the numbers of known plants increased, it becomes necessary to systemize them by using common characteristics. Indigenous societies extensive use plants to sustain life enabling them to create their own traditional naming and classification systems called ethnotaxonomy. The aim of ethnobotany is to record the indigenous knowledge of plants including ethnotaxonomic approach and their utilization by the inhabitants of the area in question. Due to industrialization, the traditional cultures and botanical wisdom is disappearing at a rapid pace. Therefore, present study was aimed to record the ethnotaxonomic perspective together with ethnobotanical uses of plants by the inhabitants of Tehsil Kotli Sattian, district Rawalpindi. The present study revealed that natives have good knowledge about the utilization of plant wealth and they are fulfilling their daily life requirement from available plant species in surrounding areas. Of them, most of the species were culturally named (157 spp.) that reflects their attachment with the native flora of the study area.

**Table 8. List of plant species having multifarious uses recorded from the study area**

| S. No. | Botanical name | Family | Multifarious Uses |
|---|---|---|---|
| 1 | *Acacia modesta* Wall. | Mimosaceae | Plant wood is very strong and used for construction proposes and furniture material. |
| 2 | *Adhatoda zeylanica* Medic | Acanthaceae | Flower gives sweetness which is used in sweet dishes. Unripe bananas are rolled in the leaves of this plant for ripening. |
| 3 | *Ailanthus altissima* (Mill.) Swingle | Simarubaceae | The straight branches are used in shattering of building. |

| S. No. | Botanical name | Family | Multifarious Uses |
|---|---|---|---|
| 4 | *Ajuga bracteosa* Wall. ex Bth. | Lamiaceae | It is cooked as vegetable along with *Rumex dentatus*. |
| 5 | *Albizzia lebbeck* (L.) Benth. | Mimosaceae | Wood of the plant is very strong is used for construction and furniture purpose. |
| 6 | *Amaranthus viridis* L. | Amaranthaceae | The plant is cooked as vegetable. |
| 7 | *Arundo donax* L. | Poaceae | Inflorescent and leaves of this plant used for making brooms. Plant stem is used for making supporting materials/roof thatching and making pencils (*Qalam*). |
| 8 | *Berberis lycium* Royle. | Berberidaceae | Leaves are used for making *Chatni*. |
| 9 | *Bombax ceiba* L. | Bombacaceae | The silky floss of seeds is used for stiffing pillows and quills. |
| 10 | *Carissa opaca* Stapf ex. Haines | Apocynaceae | Berries are used as fruit. |
| 11 | *Chenopodium album* L. | Chenopodiaceae | Plant is used as a vegetable. |
| 12 | *Chenopodium ambrosioides* L. | Chenopodiaceae | Plant is used as a vegetable. |
| 13 | *Coniogramme rosthornii* Hieron. | Coniogrammaceae | The plant is used as fodder also as ornamental. |
| 14 | *Debregeasia salicifolia* (D. Don) | Urticaceae | Branches are used in roof thatching because its wood is insect resistant. |
| 15 | *Desmostachya bipinnata* (L.) Stapf | Poaceae | Plant is used as fodder/forage. The culm is sued as broom making. Leaves are woven to make ropes. |
| 16 | *Digera muricata* (L.) Mart | Aizoaceae | The plant is eaten as vegetable by the people. |
| 17 | *Diospyros lotus* L. | Ebenaceae | Fruits are eaten as raw due to sour taste. |
| 18 | *Dodonaea viscosa* (L.) Jacq. | Sapindaceae | Plant mostly used as hedge bush. The branches are used in thatching of roof. |
| 19 | *Echinochloa crus-galli* (L.) P. Beauv. | Poaceae | Seeds were grinded and used for bread making during past time. |
| 20 | *Ficus carica* L. | Moraceae | It fruit is edible. Young leaves are cooked as vegetable. Wood is used as roof thatching. |
| 21 | *Ficus palmata* Forssk. | Moraceae | It fruit is edible. Wood is used as building material. |
| 22 | *Ficus roxburghii* Wall. ex Brand. | Moraceae | It fruit is edible. Wood is used as building material. |
| 23 | *Flacourtia indica* (Burm. F.) Merrill | Flacourtiaceae | Its fruits are edible. The wood is used in building material. The plant is used as fuel after drying. |
| 24 | *Foeniculum vulgare* Miller | Umbelliferae | Plant extract is used in as flavoring agent, spices and different beverages. |

## Table 8. (Continued)

| S. No. | Botanical name | Family | Multifarious Uses |
|---|---|---|---|
| 25 | *Grewia optiva* Drum. ex Burret. | Tiliaceae | Twigs are soaked in water for some days. These are then beaten to remove bark which is used in making ropes. |
| 26 | *Imperata cylindrica* (L) Raeuschel | Poaceae | The cottony floss is used for stuffing pillows. |
| 27 | *Mallotus philipensis* (Lam.) Muell. | Euphorbiaceae | Its wood is supposed having strong fiber and used for making furniture. |
| 28 | *Maytenus royleanus* (Wall. ex Lawson) Cufodontis | Celastraceae | The branches are used to make walking sticks |
| 29 | *Medicago denticulata* Willd. | Fabaceae | Plant is used as potherb during juvenile stage |
| 30 | *Medicago polymorpha* L. | Fabaceae | People used it as vegetable |
| 31 | *Melia azedarach* L. | Meliacea | Leaves of plant used in rice field (*Paneeri*) acting as anti-feedant. The same is also applied in storage place of wheat and rice to prevent from insect. |
| 32 | *Melilotus indica* (L.) All. | Fabaceae | Plant is cooked as potherb. |
| 33 | *Mentha longifolia* (L.) Huds. | Lamiaceae | Plant is used for flavoring agent and as condiment. |
| 34 | *Micromeria biflora* (Ham.) Bth. | Lamiaceae | The plant is rubbed in earthen pot as cleansing agent for obtaining high yield of butter. |
| 35 | *Morus alba* L. | Solanaceae | Fruits are edible. Wood is durable and used for furniture making. |
| 36 | *Morus nigra* L. | Solanaceae | Fruit is very delicious and eaten raw. Wood is used for construction propose. |
| 37 | *Myrsine africana* L. | Myrsinaceae | Fruits are edible. |
| 38 | *Nasturtim officinale* R. Br. | Brassicaceae | Leaves are cooked as vegetable. |
| 39 | *Nerium oleander* L. | | Plant is growing as an ornamental. |
| 40 | *Nyctanthes arbor-tristis* L. | Oleaceae | The plant possesses aromatic and fragrant flowers. |
| 41 | *Olea europaea* L. | Oleaceae | The wood of this plant is very strong and used for construction proposes like roof thatching and furniture making. |
| 42 | *Olea ferruginea* L. | Oleaceae | Dried leaves are used for tea making. |
| 43 | *Opuntia monacantha* Haw. | Cactaceae | Cultivated as a hedge row. |
| 44 | *Origanum vaulgare* L. | Lamiaceae | The plant is used for cleaning earthen pots for obtaining high yield of butter. |
| 45 | *Oxalis corniculata* L. | Oxalidaceae | Plant is cooked as vegetable. |

| S. No. | Botanical name | Family | Multifarious Uses |
|---|---|---|---|
| 46 | *Periploca aphylla* Decne. | Asclepiadaceae | Stem fibers are used for making ropes, thatching huts. Flowers buds are rarely eaten as raw. |
| 47 | *Phyllanthus emblica* L. | Euphorbiaceae | Wood is used for furniture making and construction. Leaves soaked in water along with soapnut (*Sapindus mukrosoi*) and Shikakai (*Acacia concinna*) and applied on hairs for long and healthy hairs. |
| 48 | *Physalis minima* L. | Poaceae | Berries are eaten as will fruit. |
| 49 | *Pinus roxburghii* Sargent | Pinaceae | Resin (*Ganda beroza*) is used in paintings. Seeds are edible. Dry leaves are used for thatching of roofs. The braches and wood is used for fuel source. Leaves supposed to have allelopathic activities. |
| 50 | *Polygonum plebejum* R. Br. | Polygonaceae | Plant used as fodder. This plant usually grows in hilly and sandy areas and acts as soil binder. |
| 51 | *Pongamia pinnata* (L.) Pierre | Papilionaceae | It is found growing as beautiful roadside tree. The wood is used in Furniture. |
| 52 | *Populus deltoides* Bartram ex Marsh. | Salicaceae | Braches are used for making agricultural tools and construction (Shattering) material. Foliage are favorite fodder to cattle |
| 53 | *Punica granatum* L. | Punicaceae | Seeds are used as flavoring agent in cooking food material. |
| 54 | *Pyrus pashia* Ham. ex D. Don | Rosaceae | The wood is used for making sticks and as fuel. Ripened fruit is eaten as raw. |
| 55 | *Quercus dilatata* Lindl. | Fagaceae | Wood is used as furniture and fuel purposes. |
| 56 | *Quercus incana* Roxb. | Fagaceae | Wood is used as furniture. |
| 57 | *Quercus leucotrichophora* A. Camus | Fagaceae | The wood is used for construction. |
| 58 | *Reinwardtia indica* Dumort. | Linaceae | The plant possesses aromatic and showy flowers and supposed to have ornamental values. |
| 59 | *Rosa brunonii* Lindll. | Rosaceae | This plant is growing for the beautification around settlements. It is also growing as hedge. |
| 60 | *Rumex hastatus* D. Don | Polygonaceae | This plant is used for making *Chatni.* Flower is used for Shabbat during summer, it has cooling effect. |

**Table 8. (Continued)**

| S. No. | ***Botanical name*** | **Family** | **Multifarious Uses** |
|---|---|---|---|
| 61 | *Saccharum bengalense* Retz. | Poaceae | The leaves are woven to make matrices. The stem/culm is used for making matrices, sitting chairs, etc. (*Patlan, Chick, Moorhe*) and other decoration pieces (e.g. cage, boxes, etc). Inflorescence is used for making brooms (*Jharoo*). Plant stem is used for making pencils (*Qalam*) and supporting materials. The culm is used for thatching of roofs as well as making *Chiks*, chairs (*Moorhe*), and decoration |
| 62 | *Saccharum spontaneum* L. | Poaceae | The leaves are woven to make matrices. The stem/culm is used for making matrices, sitting chairs, etc. (Patlan, chick, Moorhe) and other decoration pieces (e.g. cage, boxes, etc.). Inflorescence is used for making brooms (*Jharoo*). Plant stem is used for making pencils (*Qalam*) and supporting materials. The culm is used for thatching of roofs as well as making *Chiks*, chairs (*Moorhe*), and decoration. |
| 63 | *Salvia plebeia* R. Br. | Lamiaceae | The mucilaginous seeds are employed to make hair glossy. |
| 64 | *Setaria glauca* (L.) P. Beauv | Poaceae | The seeds were used for making bread in old times during famine period. |
| 65 | *Solanum nigrum* L. | Solanaceae | Its fruit is edible. Berries are eaten raw as fruit. |
| 66 | *Solanum pseudocapsicum* L. | Solanaceae | It is mostly grown as beautification purposes. It is supposed as poisonous. |
| 67 | *Stellaria media* (L.) Cyr | Celastraceae | The young twigs are cooked as vegetable. |
| 68 | *Tagetes minuta* L. | Asteraceae | The plant is also grown as ornamental. |
| 69 | *Taraxacum officinale* Weber. | Asteraceae | Leaves are cooked as vegetable. |

| S. No. | ***Botanical name*** | **Family** | **Multifarious Uses** |
|---|---|---|---|
| 70 | *Themeda anathera* (Nees) Hack. | Poaceae | Leaves are used for thatching of roof. Plant in fresh form is used as favorite fodder for cattle. The same is also collected and stored for autumn/winter season. The plant is good soil binder. Culms are used in broom making. |
| 71 | *Verbascum thapsus* L. | Scrophulariaceae | Leaves are used as substitute of cigarette (*Hukka*). |
| 72 | *Vitex negundo* L. | Verbenaceae | The plant is grown along water channels to check erosion. Leaves are used for grains storage supposed to possess insecticidal properties. The young shoots are used basket making as well as tooth brush for cleaning teeth. |
| 73 | *Zanthoxylum alatum* Roxb. | Rutaceae | Seeds are used as substitute of black pepper (*Piper nigrum*) and for making *Chatni*. |
| 74 | *Zizyphus mauritiana* Mill. | Rhamnaceae | Fruits known as *Ber* are edible. Leaves are used as fodder. Its wood is very strong and used in furniture and in construction. |
| 75 | *Zizyphus oxyphylla* Edgew. | Rhamnaceae | Fruit are edible having sour taste. |

## 4.6. Use Value Index (UVI) and Family Importance Value (FIV)

Ethnobotanical studies extracted valuable information about multifarious uses of plant species from different areas of the world. The basic need of human being was food shelter, clothing and medicines acquired from plant resources. The present study reveals that there were 225 plant species belonging to 145 genera and 67 families which served to the natives for fulfilling daily life requirement. The documented plants were used by the natives in seven major use classes in the study area (Table 9, Figure 6). The highest use of such species were in feeding livestock of the dwellers (204 spp., 90.67%), followed by medicinal use (109 spp., 48.44%), fuel (70 spp., 31.11%), others (65 spp., 28.89%), vegetable cooking (22 spp., 9.78%), wild fruits (16 spp., 7.11%) and ethno-veterinary use (4 spp., 1.78%).

The use index of species indicated that one species (*Ficus acrica*) was found as heavily used which accomplished sex use categories/needs out of seven, whereas, nine species such as *Berberis lyceum, Flacourtia indica, Morus alba, M. nigra, Punica granatum, Pyrus pashia, Zanthoxylum alatum, Zizyphus mauritiana* and *Z. oxyphylla* fulfilled 5 use categories of the local communities (Table 7). Twenty five species were used for the accomplishment of four use categories, 39 species in three categories, 71 species in two use classes, whereas, 81 species were singly used.

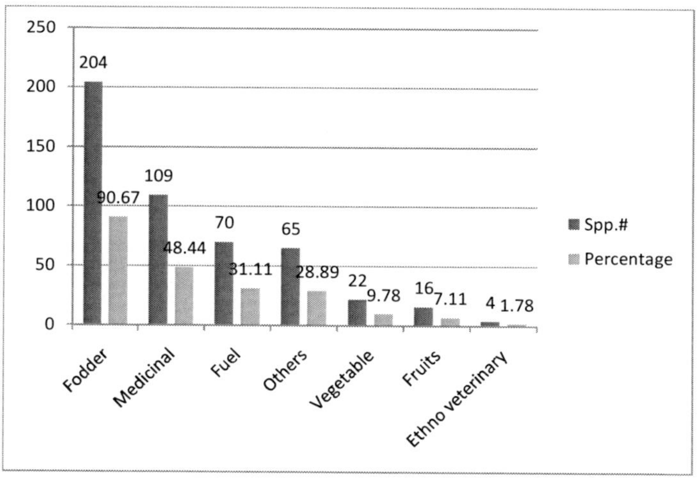


Figure 6. Graphical presentation of plants used I different use classes.

**Table 9. Use value index (UVI) for all plant species recorded from the study area**

| Sr# | Plant Name | Family | Fruit | Vegetable | Medicinal | Ethno veterinary | Fuel | Fodder | Other | Usage Index |
|---|---|---|---|---|---|---|---|---|---|---|
| 1 | *Ficus carica* L. | Moraceae | + | + | + | — | + | + | + | 6 |
| 2 | *Berberis lycium* Royle. | Berberidaceae | — | + | + | — | + | + | + | 5 |
| 3 | *Flacourtia indica* (Burm. f.) Merr. | Flacourtiaceae | + | — | + | — | + | + | + | 5 |
| 4 | *Morus alba* L. | Moraceae | + | — | + | — | + | + | + | 5 |
| 5 | *Morus nigra* L. | Moraceae | + | — | + | — | + | + | + | 5 |
| 6 | *Punica granatum* L. | Punicaceae | — | + | + | — | + | + | + | 5 |
| 7 | *Pyrus pashia* | Rosaceae | + | — | + | — | + | + | + | 5 |

| Sr# | Plant Name | Family | Fruit | Vegetable | Medicinal | Ethno veterinary | Fuel | Fodder | Other | Usage Index |
|---|---|---|---|---|---|---|---|---|---|---|
| 8 | *Zanthoxylum alatum* Roxb. | Rutaceae | – | + | + | – | + | + | + | 5 |
| 9 | *Zizyphus mauritiana* Mill. | Rhamnaceae | + | – | + | – | + | + | + | 5 |
| 10 | *Zizyphus oxyphylla* Edgew. | Rhamnaceae | + | – | + | – | + | + | + | 5 |
| 11 | *Acacia modesta* Wall. | Mimosaceae | – | – | + | – | + | + | + | 4 |
| 12 | *Albizzia lebbeck* (L.) Benth. | Mimosaceae | – | – | + | – | + | + | + | 4 |
| 13 | *Bombax ceiba* L. | Bombacaceae | – | – | + | – | + | + | + | 4 |
| 14 | *Carissa opaca* Stapf ex. Haines | Apocynaceae | + | – | + | – | + | + | – | 4 |
| 15 | *Debregeasia salicifolia* (D. Don) Rendle | Urticaceae | – | – | + | – | + | + | + | 4 |
| 16 | *Diospyros lotus* L. | Ebenaceae | + | – | + | – | + | + | – | 4 |
| 17 | *Eruca sativa* L. | Brassicaceae | – | – | + | + | + | + | – | 4 |
| 18 | *Ficus religiosa* L. | Moraceae | + | – | – | – | + | + | + | 4 |
| 18 | *Ficus roxburghii* Wall. ex Brand. | Moraceae | – | – | + | – | + | + | + | 4 |
| 19 | *Foeniculum vulgare* (Miller) | Apiaceae | – | – | + | – | + | + | + | 4 |
| 20 | *Grewia optiva* Drum. ex. Burret. | Tiliaceae | – | – | + | – | + | + | + | 4 |
| 21 | *Mallotus philipensis* (Lam.) Muell. | Euphorbiaceae | – | – | + | – | + | + | + | 4 |
| 22 | *Myrsine africana* L. | Myrsinaceae | + | – | + | – | + | + | – | 4 |
| 23 | *Olea europaea* L. | Oleaceae | – | – | + | – | + | + | + | 4 |
| 24 | *Olea ferruginea* Royle | Oleaceae | – | – | + | – | + | + | + | 4 |
| 25 | *Phyllanthus emblica* L. | Euphorbiaceae | – | – | + | – | + | + | + | 4 |
| 26 | *Physalis divaricata* D.Don. | Solanaceae | + | – | + | – | + | + | – | 4 |
| 27 | *Pinus roxburghii* Sargent | Pinaceae | – | – | + | – | + | + | + | 4 |
| 28 | *Pongamia pinnata*(L.) Pierre | Fabaceae | – | – | + | – | + | + | + | 4 |
| 29 | *Quercus leucotrichophora* | Fagaceae | – | – | + | – | + | + | + | 4 |
| 30 | *Rosa brunonii* Lindll. | Rosaceae | – | – | + | – | + | + | + | 4 |
| 31 | *Rubus ellipticus* Smith | Rubiaceae | + | – | + | – | + | + | – | 4 |
| 32 | *Rumex hastatus* D. Don. | Polygonaceae | – | + | + | – | – | + | + | 4 |
| 33 | *Verbascum thapsus* L. | Scrophulariaceae | – | – | + | – | + | + | + | 4 |
| 34 | *Vitex negundo* L. | Verbenaceae | – | – | + | – | + | + | + | 4 |
| 35 | *Abutilon bidentatum* Hochst. ex Rich. | Malvaceae | – | – | + | – | + | + | – | 3 |
| 36 | *Ailanthus altissima* (Mill.) Swingle | Simarubaceae | – | – | – | – | + | + | + | 3 |
| 37 | *Ajuga bracteosa* Wall. ex Bth. | Lamiaceae | – | + | + | – | – | + | – | 3 |

## Table 9. (Continued)

| Sr# | *Plant Name* | Family | Fruit | Vegetable | Medicinal | Ethno veterinary | Fuel | Fodder | Other | Usage Index |
|---|---|---|---|---|---|---|---|---|---|---|
| 38 | *Arundo donax* L. | Poaceae | — | — | + | — | — | + | + | 3 |
| 39 | *Boerhavia procumbens* Banks ex Roxb. | Nyctaginaceae | — | + | — | — | — | + | + | 3 |
| 40 | *Calotropis procera* (Willd.) R.Br. | Asclepiadaceae | — | — | + | — | + | + | — | 3 |
| 41 | *Chenopodium album* L. | Chenopodiaceae | — | + | + | — | — | + | — | 3 |
| 42 | *Chenopodium ambrosioides* L. | Chenopodiaceae | — | + | + | — | — | + | — | 3 |
| 43 | *Colebrookia oppositifolia* Sm. | Lamiaceae | — | — | + | — | + | + | — | 3 |
| 44 | *Datura stramonium* L. | Solanaceae | — | — | + | — | + | + | — | 3 |
| 45 | *Dodonaea viscosa* (L.) Jacq. | Sapindaceae | — | — | — | — | + | + | + | 3 |
| 46 | *Ficus bengalensis* L. | Moraceae | — | — | + | — | + | + | — | 3 |
| 47 | *Ficus palmata* Forssk. | Moraceae | — | — | — | — | + | + | + | 3 |
| 48 | *Hedera nepallensis* K. Koch | Araliaceae | — | — | + | — | + | + | — | 3 |
| 49 | *Lantana camara* L. | Verbenaceae | — | — | — | + | + | — | + | 3 |
| 50 | *Malva neglecta* Waller. | Malvaceae | + | — | + | — | — | — | + | 3 |
| 51 | *Maytenus royleanus* (Wall. ex Lawson) Cufodontis | Celastraceae | — | — | — | — | + | + | + | 3 |
| 52 | *Mentha longifolia* (L.) Huds | Lamiaceae | — | — | + | — | — | + | + | 3 |
| 53 | *Nyctanthes arbor-tristis* L. | Oleaceae | — | — | + | — | — | + | + | 3 |
| 54 | *Oxalis corniculata* L. | Oxalidaceae | — | + | + | — | — | + | — | 3 |
| 55 | *Pistacia chinensis Bunge* | Anacardiaceae | — | — | + | — | + | + | — | 3 |
| 56 | *Polygonum plebejum* R. Br. | Polygonaceae | — | — | + | — | — | + | + | 3 |
| 57 | *Quercus dilatata* Lindley | Fagaceae | — | — | — | — | + | + | + | 3 |
| 58 | *Quercus incana* Roxb. | Fagaceae | — | — | — | — | + | + | + | 3 |
| 59 | *Rumex dentatus* L. | Polygonaceae | — | + | + | — | — | + | — | 3 |
| 60 | *Saccharum bengalense* Retz. | Poaceae | — | — | — | — | + | + | + | 3 |
| 61 | *Saccharum spontaneum* L. | Poaceae | — | — | — | — | + | + | + | 3 |
| 62 | *Sageretia thea* (Osback) M. C. Johnston | Rhamnaceae | + | — | — | — | + | + | — | 3 |
| 63 | *Salvia plebeia R. Br.* | Lamiaceae | — | — | + | — | — | + | + | 3 |
| 64 | *Silybum marianum* (L.) Gaertin | Asteraceae | — | — | + | — | + | + | — | 3 |
| 65 | *Solanum incanum* L. | Solanaceae | — | + | + | — | — | + | — | 3 |
| 66 | *Solanum nigrum* L. | Solanaceae | — | + | + | — | — | + | — | 3 |

| Sr# | Plant Name | Family | Fruit | Vegetable | Medicinal | Ethno veterinary | Fuel | Fodder | Other | Usage Index |
|---|---|---|---|---|---|---|---|---|---|---|
| 67 | *Solanum surattense* Burm.f. | Solanaceae | – | – | + | – | + | + | – | 3 |
| 68 | *Solanum villosum* (L.) Moench | Solanaceae | + | – | + | – | – | + | – | 3 |
| 69 | *Sorghum halepense* (L.) Bern. | Poaceae | – | – | – | + | – | + | + | 3 |
| 70 | *Taraxacum officinale* Weber. | Asteraceae | – | + | – | – | + | + | – | 3 |
| 71 | *Themeda anathera* (Nees) Hack | Poaceae | – | – | – | – | + | + | + | 3 |
| 72 | *Withania somnifera* (L.) Dunal. | Solanaceae | – | – | + | – | + | + | – | 3 |
| 73 | *Woodfordia fruticosa* (L.) S. Kurz | Vitaceae | – | – | + | – | + | + | – | 3 |
| 74 | *Achyranthus aspera* L. | Amaranthaceae | – | – | + | – | – | + | – | 2 |
| 75 | *Adhatoda zeylanica* Medic. | Acanthaceae | – | – | + | – | – | – | + | 2 |
| 76 | *Adiantum capillus-veneris* L. | Adiantaceae | – | – | + | – | – | + | – | 2 |
| 77 | *Adiantum incissum* | Adiantaceae | – | – | + | – | – | + | – | 2 |
| 78 | *Alternanthera pungens* Kunth. | Amaranthaceae | – | – | + | – | – | + | – | 2 |
| 79 | *Amaranthus hybridus* L. | Amaranthaceae | – | – | + | – | – | + | – | 2 |
| 80 | *Amaranthus viridis* L. | Amaranthaceae | – | – | + | – | – | + | – | 2 |
| 81 | *Asphodelus tenuifolius* Cavan. | Liliaceae | – | + | – | – | – | + | – | 2 |
| 82 | *Astragalus squarrosus* Bunge | Fabaceae | – | – | – | – | + | + | – | 2 |
| 83 | *Bacopa monnieri* (L.) Wettst. | Scrophulariaceae | – | – | + | – | – | + | – | 2 |
| 84 | *Cannabis sativa* L. | Cannabinaceae | – | – | – | + | – | + | – | 2 |
| 85 | *Cardiospermum halicacabum* L. | Sapindaceae | – | – | + | – | – | + | – | 2 |
| 86 | *Carthamus oxycantha* M. Bieb | Asteraceae | – | – | – | – | + | + | – | 2 |
| 87 | *Chrysopogon aucheri* (Boiss.) Stapf. | Poaceae | – | – | + | – | – | + | – | 2 |
| 88 | *Clematis montana* Buch. | Rananculaceae | – | – | – | – | + | + | – | 2 |
| 89 | *Colchicum aitchisonii* (Hook. f.) E. Nasir | Liliaceae | – | – | + | – | – | + | – | 2 |
| 90 | *Coniogramme rosthornii* Hieron. | Coniogrammaceae | – | – | – | – | – | + | + | 2 |
| 91 | *Convolvulus arvensis* L. | Convulvulaceae | – | + | – | – | – | + | – | 2 |
| 92 | *Cuscuta reflexa* Roxb. | Cuscutaceae | – | – | + | – | – | + | – | 2 |

## Table 9. (Continued)

| Sr# | Plant Name | Family | Fruit | Vegetable | Medicinal | Ethno veterinary | Fuel | Fodder | Other | Usage Index |
|---|---|---|---|---|---|---|---|---|---|---|
| 93 | *Cynodon dactylon* L. Pers | Poaceae | — | — | + | — | — | + | — | 2 |
| 94 | *Cyperus rotundus* L. | Cyperaceae | — | — | + | — | — | + | — | 2 |
| 95 | *Desmostachya bipinnata* (L.) Stapf | Poaceae | — | — | — | — | — | + | + | 2 |
| 96 | *Dicanthium annulatum* (Forssk.) Stapf | Poaceae | — | — | — | — | + | + | — | 2 |
| 97 | *Dioscorea deltoidea* Wall. ex Kunth | Dioscoreaceae | — | — | — | — | + | + | — | 2 |
| 98 | *Echinochloa crus-galli* (L.) P. Beauv. | Poaceae | — | — | — | — | — | + | + | 2 |
| 99 | *Heliotopium europaeum* L. | Boraginaceae | — | — | + | — | — | + | — | 2 |
| 100 | *Indigofera himalayensis* Ali | Fabaceae | — | — | — | — | + | + | — | 2 |
| 101 | *Ipomoea carnea* Jacq. | Convolvulaceae | — | — | + | — | + | — | — | 2 |
| 102 | *Lannea coromandelica* L. | Anacardiaceae | — | — | + | — | — | + | — | 2 |
| 103 | *Malvastrum coromendelianum* L. | Malvaceae | — | — | — | — | — | + | + | 2 |
| 104 | *Medicago denticulata* Willd. | Fabaceae | — | + | — | — | — | + | — | 2 |
| 105 | *Medicago laciniata* (L.) Mill. | Fabaceae | — | + | — | — | — | + | — | 2 |
| 106 | *Medicago polymorpha* L. | Fabaceae | — | + | — | — | — | + | — | 2 |
| 107 | *Melia azedarach* L. | Meliacea | — | — | + | — | — | — | + | 2 |
| 108 | *Mentha royleana* Wall. ex Bth. | Lamiaceae | — | — | + | — | — | — | + | 2 |
| 109 | *Micromeria biflora* (Ham.) Bth. | Lamiaceae | — | — | — | — | — | + | + | 2 |
| 110 | *Mimosa rubicaulis* Lam. | Mimosaceae | — | — | + | — | — | + | — | 2 |
| 111 | *Nerium oleander* L. | Apocynaceae | — | — | + | — | — | — | + | 2 |
| 112 | *Opuntia monacantha* Haw. | Cactaceae | — | — | + | — | — | — | + | 2 |
| 113 | *Origanum vaulgare* L. | Lamiaceae | — | — | + | — | — | — | + | 2 |
| 114 | *Otostegia limbata* (Benth.) Boiss. | Lamiaceae | — | — | + | — | — | + | — | 2 |
| 115 | *Peganum harmala* L. | Zygophyllaceae | — | — | + | — | — | + | — | 2 |
| 116 | *Periploca aphylla* Decne. | Asclepiadaceae | — | — | + | — | — | — | + | 2 |
| 117 | *Phyla nodiflora* (L.) Greene | Verbenaceae | — | — | + | — | — | + | — | 2 |
| 118 | *Plantago lanceolata* L. | Plantaginaceae | — | — | + | — | — | + | — | 2 |
| 119 | *Plantago major* L. | Plantaginaceae | — | — | + | — | — | + | — | 2 |
| 120 | *Plantago ovata* Frossk. | Plantaginaceae | — | — | + | — | — | + | — | 2 |

| Sr# | Plant Name | Family | Fruit | Vegetable | Medicinal | Ethno veterinary | Fuel | Fodder | Other | Usage Index |
|---|---|---|---|---|---|---|---|---|---|---|
| 121 | *Populus deltoides* Bartram ex Marsh. | Salicaceae | – | – | – | – | + | + | – | 2 |
| 122 | *Potentilla supina* L. | Rosaceae | – | – | + | – | – | + | – | 2 |
| 123 | *Pueraria tuberosa* (Roxb. Ex Willd) DC. | Fabaceae | – | – | + | – | – | + | – | 2 |
| 124 | *Rananculus muricatus* L. | Rananculaceae | – | – | + | – | – | + | – | 2 |
| 125 | *Ranunculus arvensis* L. | Ranunculaceae | – | – | + | – | – | + | – | 2 |
| 126 | *Reinwardtia indica* Dumort. | Linaceae | – | – | + | – | – | – | + | 2 |
| 127 | *Ricinis communis* L. | Euphorbiaceae | – | – | + | – | + | – | – | 2 |
| 128 | *Rubia cordifolia* L. | Rubiaceae | – | – | + | – | – | + | – | 2 |
| 129 | *Rumex nepalensis* Spreng | Polygonaceae | – | – | + | – | – | + | – | 2 |
| 130 | *Salvia moorcroftiana* Wall. ex Bth. | Lamiaceae | – | – | + | – | – | + | – | 2 |
| 131 | *Sauromatum venosum* (Ait.) Schott. | Araceae | – | + | – | – | – | + | – | 2 |
| 132 | *Setaria glauca* (L) P.Beauv | Poaceae | – | – | – | – | – | + | + | 2 |
| 133 | *Sida cordata* (Burm. f.) Borss.-Waalkes | Malvaceae | – | – | + | – | – | + | – | 2 |
| 134 | *Sisymbrium irio* L. | Brassicaceae | – | – | + | – | – | + | – | 2 |
| 135 | *Smilax aspera* L. | Smilacaceae | – | – | + | – | – | + | – | 2 |
| 136 | *Stellaria media* (L.) Cry. | Caryophyllaceae | – | + | – | – | – | + | – | 2 |
| 137 | *Tagetes minuta* L. | Asteraceae | – | – | + | – | – | – | + | 2 |
| 138 | *Tribulus terrestris* L. | Zygophyllaceae | – | – | + | – | – | + | – | 2 |
| 139 | *Trichodesma indicum* (L.) R. Br. | Boraginaceae | – | – | + | – | – | + | – | 2 |
| 140 | *Urtica dioica* L. | Urticaceae | – | – | + | – | – | – | + | 2 |
| 141 | *Valeriana wallichii* DC. | Valerianaceae | – | – | + | – | – | + | – | 2 |
| 142 | *Verbena officinalis* L. | Verbenaceae | – | – | + | – | – | + | – | 2 |
| 143 | *Vicia sativa* L. | Fabaceae | – | – | + | – | – | + | – | 2 |
| 144 | *Viola cancescens* Wall. ex Roxb. | Violaceae | – | – | + | – | – | + | – | 2 |
| 145 | *Ajuga parviflora* Bth. | Lamiaceae | – | – | – | – | – | + | – | 1 |
| 146 | *Amaranthus spinosus* L. | Amaranthaceae | – | – | – | – | – | + | – | 1 |
| 147 | *Anagalis arvensis* L. | Primulaceae | – | – | – | – | – | + | – | 1 |
| 148 | *Arabis himalaica* (Edgew.) O.E. Schulz | Brassicaceae | – | – | – | – | – | + | – | 1 |
| 150 | *Argyrolobium helleborifolium* Sonott. | Fabaceae | – | – | – | – | – | + | – | 1 |
| 149 | *Argyrolobium roseum* (Camb.) Jaub & Spach | Fabaceae | – | – | – | – | + | – | – | 1 |
| 151 | *Aristida cyanatha* Nees ex Steud. | Poaceae | – | – | – | – | – | + | – | 1 |

**Table 9. (Continued)**

| Sr# | Plant Name | Family | Fruit | Vegetable | Medicinal | Ethno veterinary | Fuel | Fodder | Other | Usage Index |
|---|---|---|---|---|---|---|---|---|---|---|
| 152 | *Artemisia scoparia* Waldst. & Kit. | Asteraceae | – | – | – | – | – | + | – | 1 |
| 153 | *Asparagus gracilis* Royle | Asparagaceae | – | – | – | – | + | – | – | 1 |
| 154 | *Avena fatua* L. | Poaceae | – | – | – | – | – | + | – | 1 |
| 155 | *Barleria acanthoides* Vahl | Acanthaceae | – | – | – | – | – | + | – | 1 |
| 156 | *Barleria cristata* L. | Acanthaceae | – | – | – | – | – | + | – | 1 |
| 157 | *Capsella bursa-pastoris* (L.) Medik | Brassicaceae | – | – | – | – | – | + | – | 1 |
| 158 | *Cenchrus ciliaris* L. | Poaceae | – | – | – | – | – | + | – | 1 |
| 159 | *Cenchrus pennisetiformis* Hochst. & Steud. | Poaceae | – | – | – | – | – | + | – | 1 |
| 160 | *Cenchrus setigerus* Vahl | Poaceae | – | – | – | – | – | + | – | 1 |
| 161 | *Cirsium arvense* (L.) Scope. | Asteraceae | – | – | – | – | – | + | – | 1 |
| 162 | *Clematis napaulensis* Royle | Rananculaceae | – | – | – | – | – | + | – | 1 |
| 163 | *Commelina paludosa* Blume | Commelinaceae | – | – | – | – | – | + | – | 1 |
| 164 | *Conyza aegyptica* Ait. | Asteraceae | – | – | – | – | – | + | – | 1 |
| 165 | *Conyza bonariensis* L. | Asteraceae | – | – | – | – | – | + | – | 1 |
| 166 | *Conyza canadensis* L. | Asteraceae | – | – | – | – | – | + | – | 1 |
| 167 | *Coronopus didymus* (L.) Sm. | Brassicaceae | – | – | – | – | – | + | – | 1 |
| 168 | *Crotolaria medicagnea* Lam. | Fabaceae | – | – | – | – | – | + | – | 1 |
| 169 | *Dactyloctenium aegyptium* L. | Poaceae | – | – | – | – | – | + | – | 1 |
| 170 | *Dicliptera roxburghiana* Nees | Acanthaceae | – | – | – | – | – | + | – | 1 |
| 171 | *Digera muricata* (L.) Mart. | Amaranthaceae | – | – | – | – | – | + | – | 1 |
| 172 | *Echinops echinatus* Roxb. | Asteraceae | – | – | – | – | – | + | – | 1 |
| 173 | *Eclipta prostrata* (L.) L. | Asteraceae | – | – | – | – | – | + | – | 1 |
| 174 | *Equisetum ramosissimum* (Desf.) | Equistaceae | – | – | – | – | – | – | + | 1 |
| 175 | *Eragrostis ateroviens* (Desf.) Trin. ex Nees | Poaceae | – | – | – | – | – | + | – | 1 |
| 176 | *Eragrostis minor* Host. | Poaceae | – | – | – | – | – | + | – | 1 |
| 177 | *Euphorbia clarkeana* Hkf. | Euphorbiaceae | – | – | – | – | – | + | – | 1 |
| 178 | *Euphorbia granulata* Forssk. | Euphorbiaceae | – | – | – | – | – | + | – | 1 |

| Sr# | *Plant Name* | Family | Fruit | Vegetable | Medicinal | Ethno veterinary | Fuel | Fodder | Other | Usage Index |
|---|---|---|---|---|---|---|---|---|---|---|
| 179 | *Euphorbia helioscopia* Mewski. | Euphorbiaceae | – | – | – | – | – | + | – | 1 |
| 180 | *Euphorbia hirta* L. | Euphorbiaceae | – | – | – | – | – | + | – | 1 |
| 181 | *Euphorbia indica* Lam. | Euphorbiaceae | – | – | – | – | – | + | – | 1 |
| 182 | *Euphorbia prostrata (*L.) Ait | Euphorbiaceae | – | – | – | – | – | + | – | 1 |
| 183 | *Euphrasia himalayica* Wettst. | Scrophulariaceae | – | – | – | – | – | + | – | 1 |
| 184 | *Fumaria indica* (Hausskn.) H.N. Pugsley | Fumariaceae | – | – | – | – | – | + | – | 1 |
| 185 | *Gallium aparine* L. | Rubiaceae | – | – | – | – | – | + | – | 1 |
| 186 | *Geranium rotundifolium* L. | Geraniaceae | – | – | – | – | – | + | – | 1 |
| 187 | *Grewia tenax*(Forssk.) Aschers & Schweinf. | Tiliaceae | – | – | – | – | – | + | – | 1 |
| 188 | *Imperata cylindrica* (L.) Raeuschel | Poaceae | – | – | – | – | – | + | – | 1 |
| 189 | *Indigofera linifolia* (L. f) Retz | Fabaceae | – | – | – | – | – | + | – | 1 |
| 190 | *Indigofera sessiliflora* DC | Fabaceae | – | – | – | – | – | + | – | 1 |
| 191 | *Ipomoea hederacea* (L.) Jacq. | Convolvulaceae | – | – | – | – | – | + | – | 1 |
| 192 | *Ipomoea nil* (L.) Roth | Convolvulaceae | – | – | – | – | – | + | – | 1 |
| 193 | *Kickxia ramosissima* (Wall.) Janchen | Scrophulariaceae | – | – | – | – | – | + | – | 1 |
| 194 | *Lactuca auriculata* (Wall. ex Dc.) | Asteraceae | – | – | – | – | – | + | – | 1 |
| 195 | *Lactuca dissecta* D. Don. | Asteraceae | – | – | – | – | – | + | – | 1 |
| 196 | *Lactuca serriola* L. | Asteraceae | – | – | – | – | – | + | – | 1 |
| 197 | *Lantana indica* Roxb. | Verbenaceae | – | – | – | – | + | – | – | 1 |
| 198 | *Lathyrus aphaca* L. | Fabaceae | – | – | – | – | – | + | – | 1 |
| 199 | *Launaea procumbens* (Roxb.) Ram. & Rajgo. | Asteraceae | – | – | – | – | – | + | – | 1 |
| 200 | *Lepidium sativum* L. | Brassicaceae | – | – | – | – | – | + | – | 1 |
| 201 | *Lotus corniculatus* (Wald. & Kit. ex Willd.) Briq. & Rech. F. | Fabaceae | – | – | – | – | – | + | – | 1 |
| 225 | *Melilotus indica* (L.) All. | Fabaceae | – | – | – | – | – | + | – | 1 |
| 202 | *Nasturtim officinale* R. Br. | Brassicaceae | – | + | – | – | – | – | – | 1 |
| 203 | *Parthenium hytserophorus* L. | Asteraceae | – | – | – | – | – | + | – | 1 |
| 204 | *Phalaris minor* Retz. | Poaceae | – | – | – | – | – | + | – | 1 |
| 205 | *Physalis minima* L. | Solanaceae | – | – | – | – | – | + | – | 1 |
| 206 | *Polygonum barbatum* L. | Polygonaceae | – | – | – | – | – | + | – | 1 |

**Table 9. (Continued)**

| Sr# | *Plant Name* | Family | Fruit | Vegetable | Medicinal | Ethno veterinary | Fuel | Fodder | Other | Usage Index |
|---|---|---|---|---|---|---|---|---|---|---|
| 207 | *Polypogon fugax* Nees ex Steud. | Poaceae | – | – | – | – | – | + | – | 1 |
| 208 | *Psammogeton biternatum* Edgew. | Apiaceae | – | – | – | – | – | + | – | 1 |
| 209 | *Pteridium aquilinum* (L.) Kuhn | Pteridaceae | – | – | – | – | – | + | – | 1 |
| 210 | *Rananculus sceleratus* L. | Rananculaceae | – | – | – | – | – | + | – | 1 |
| 211 | *Rhynchosia minima* (L.) DC | Fabaceae | – | – | – | – | – | + | – | 1 |
| 212 | *Saussurea albescens* (DC.) Schr. Bip. | Asteraceae | – | – | – | – | – | + | – | 1 |
| 213 | *Saussurea atkinsonii* (Clarke) | Asteraceae | – | – | – | – | – | + | – | 1 |
| 214 | *Saussurea heteromalla* DC. | Asteraceae | – | – | – | – | – | + | – | 1 |
| 215 | *Silene conoidae* L. | Caryophyllaceae | – | – | – | – | – | + | – | 1 |
| 216 | *Solanum pseudocapsicum* L. | Solanaceae | – | – | – | – | – | – | + | 1 |
| 217 | *Sonchus arvense* | Asteraceae | – | – | – | – | – | + | – | 1 |
| 218 | *Sonchus asper* (L.) Hill. | Asteraceae | – | – | – | – | – | + | – | 1 |
| 219 | *Sorghum bicolor* (L.) Moench. | Poaceae | – | – | – | – | – | + | – | 1 |
| 220 | *Taraxcum wallichii* DC. | Asteraceae | – | – | – | – | – | + | – | 1 |
| 221 | *Trianthema portulacastrum* L. | Aizoaceae | – | – | – | – | – | + | – | 1 |
| 222 | *Trifolium repens* L. | Fabaceae | – | – | – | – | – | + | – | 1 |
| 223 | *Vicia faba* L. | Fabaceae | – | – | – | – | – | + | – | 1 |
| 224 | *Xanthium strumarium* L. | Asteraceae | – | – | – | – | + | – | – | 1 |
| | | | 16 | 22 | 109 | 4 | 70 | 205 | 65 | |

The present investigation reported that 68 plant families contributed the total ethnobotanical flora of Kotli Sattian including 3 ferns, 1 gymnosperm, 32 monocoteledons and 189 dicotyledon species (Figure 8). Asteraceae and Poaceae were found the most dominating families which contributed each 23 plants (10.18% each), followed by Fabaceae (19 spp., 8.41%), Lamiaceae (10 spp., 4.42%), Euphorbiaceae, Solanaceae (9 spp., 3.98% each), Brassicaceae, Moraceae (7 spp., 3.10% each), Amaranthaceae, Verbenaceae (6 spp., 2.65% each) and Polygonaceae & Rananculaceae ( 5 spp., 2.21% each); while, rest of the families had few number of species (Table 8).

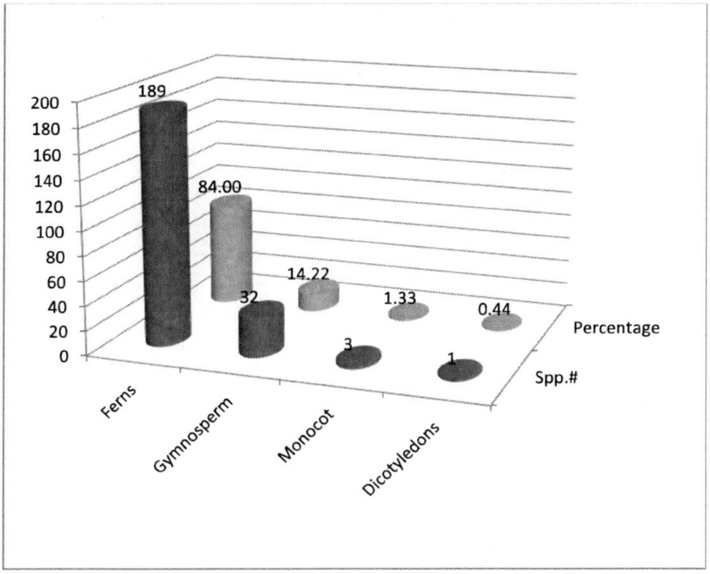


Figure 7. Proportion of the floral groups in the flora.

**Table 10. Contribution of groups/plant families in the ethnobotanical flora of Kotli Sattian**

| S. No. | Group/Family | Spp.# | Percentage |
|---|---|---|---|
| 1 | Ferns | 3 | 1.33 |
| 2 | Gymnosperm | 1 | 0.44 |
| | MONOCOTYLEDONS | | |
| 3 | Poaceae | 23 | 10.18 |
| 4 | Asparagaceae | 3 | 1.33 |
| 5 | Liliaceae | 2 | 0.88 |
| 6 | Araceae | 1 | 0.44 |
| 7 | Araliaceae | 1 | 0.44 |
| 8 | Commelinaceae | 1 | 0.44 |
| 9 | Cyperaceae | 1 | 0.44 |
| | DICOTYLEDONS | | |
| 10 | Asteraceae | 23 | 10.18 |
| 11 | Fabaceae | 19 | 8.41 |
| 12 | Lamiaceae | 10 | 4.42 |
| 13 | Euphorbiaceae | 9 | 3.98 |
| 14 | Solanaceae | 9 | 3.98 |
| 15 | Brassicaceae | 7 | 3.10 |
| 16 | Moraceae | 7 | 3.10 |

**Table 10. (Continued)**

| S. No. | Group/Family | Spp.# | Percentage |
|---|---|---|---|
| 17 | Amaranthaceae | 6 | 2.65 |
| 18 | Verbenaceae | 6 | 2.65 |
| 19 | Polygonaceae | 5 | 2.21 |
| 20 | Ranunculaceae | 5 | 2.21 |
| 21 | Acanthaceae | 4 | 1.77 |
| 22 | Convulvulaceae | 4 | 1.77 |
| 23 | Malvaceae | 4 | 1.77 |
| 24 | Scrophulariaceae | 4 | 1.77 |
| 25 | Fagaceae | 3 | 1.33 |
| 26 | Mimosaceae | 3 | 1.33 |
| 27 | Oleaceae | 3 | 1.33 |
| 28 | Plantaginaceae | 3 | 1.33 |
| 29 | Rhamnaceae | 3 | 1.33 |
| 30 | Rosaceae | 3 | 1.33 |
| 31 | Rubiaceae | 3 | 1.33 |
| 32 | Anacardiaceae | 2 | 0.88 |
| 33 | Apiaceae | 2 | 0.88 |
| 34 | Apocynaceae | 2 | 0.88 |
| 35 | Boraginaceae | 2 | 0.88 |
| 36 | Caryophyllaceae | 2 | 0.88 |
| 37 | Chenopodiaceae | 2 | 0.88 |
| 38 | Sapindaceae | 2 | 0.88 |
| 39 | Tiliaceae | 2 | 0.88 |
| 40 | Urticaceae | 2 | 0.88 |
| 41 | Vitaceae | 2 | 0.88 |
| 42 | Zygophyllaceae | 2 | 0.88 |
| 43 | Aizoaceae | 1 | 0.44 |
| 44 | Araliaceae | 1 | 0.44 |
| 45 | Berberidaceae | 1 | 0.44 |
| 46 | Bombacaceae | 1 | 0.44 |
| 47 | Cactaceae | 1 | 0.44 |
| 48 | Cannabinaceae | 1 | 0.44 |
| 49 | Celastraceae | 1 | 0.44 |
| 50 | Commelinaceae | 1 | 0.44 |
| 51 | Cuscutaceae | 1 | 0.44 |
| 52 | Dioscoreaceae | 1 | 0.44 |
| 53 | Ebenaceae | 1 | 0.44 |
| 54 | Flacourtiaceae | 1 | 0.44 |
| 55 | Fumariaceae | 1 | 0.44 |
| 56 | Geraniaceae | 1 | 0.44 |

| S. No. | Group/Family | Spp.# | Percentage |
|---|---|---|---|
| 57 | Linaceae | 1 | 0.44 |
| 58 | Meliaceae | 1 | 0.44 |
| 59 | Myrsinaceae | 1 | 0.44 |
| 60 | Nyctaginaceae | 1 | 0.44 |
| 61 | Oxalidaceae | 1 | 0.44 |
| 62 | Primulaceae | 1 | 0.44 |
| 63 | Punicaceae | 1 | 0.44 |
| 64 | Rutaceae | 1 | 0.44 |
| 65 | Salicaceae | 1 | 0.44 |
| 66 | Simarubaceae | 1 | 0.44 |
| 67 | Smilacaceae | 1 | 0.44 |

*Acacia modesta*

*Adhatoda zeylanica*

*Ailanthus altissima*

*Ajuga parviflora*

*Amaranthus viridis*

*Bacopa monierii*

Plate 8. Showing habit and floral parts of some native plant species.

*Berberis lyceum*

*Boerhavia procumbens*

*Calotropis procera*

*Collebrookia oppositifolia*

*Corchorus tridens*

*Echinops echinatus*

Plate 8. (Continued)

*Ficus carica*

*Grewia optiva*

Plate 8. (Continued).

*Grewia tenax*

*Indigofera linifolia*

Kickxia ramosissima

*Lantana indica*

*Launaea procumbens*

*Mallotus philipiensis*

*Malva parviflora*

*Opuntia deliniii*

Plate 8. (Continued).

*Otostegia limbata*

*Periploca aphylla*

*Pinus roxburghii*

*Punica granatum*

*Rhynchosia pulverulenta*

*Salvia moorcroftiana*

*Salvia santilinifolia*

*Segeretia theezans*

Plate 8. (Continued).

*Solanum incanum*

*Solanum nigrum*

*Tribulus terrestris*

*Trichodesma indicum*

*Woodfordia fruiticosa*

Plate 8. (Continued).

# CONCLUSION AND RECOMMENDATIONS

The findings of the present study revealed that the plants of the area are providing most of the basic requirement for the survival of the local communities. The plants of the area are used for fulfilling different needs of the local inhabitants. The majority of these plants were used for fodder/forage purpose, since livelihood of the population is mostly dependant on livestock keeping. The area has very good grazing potential and found plentiful amount of palatable species. The second use of plant species was as fuel source. Medicinal use of plants is in common practice however, there was an alarming situation that the knowledge regarding different medicinal recipes was only found amongst the elders and the rate of transferral of this knowledge to the young generation was very low. In present situation; there is need to record this valuable knowledge through conducting comprehensive surveys of the area.

Another concern is that some of these plants are becoming extinct and if not preserved would vanish, so there is need to protect these plants. One way to protect these plants is to provide alternate source of their basic needs like Sui gas to the local people as they are removing all the available plants for fire wood. Moreover, there should be some legislation passed by the government for check and balance of cutting the firewood.

The present study is first sttrmpt from the study area to not only document the plants of the area but also to record the indigenous knowledge of plants particularly recipes used for the treatment for various ailments. There is also need to authenticate the recipes documented in the research through ethnopharmacological researches.

The present research would serve as baseline information for the future researchers to plan and conduct further researches in the following area in the light of the study findings:

1. Population census of plant species for documenting dominant, abundant, rare, vulnerable and threatened species from the study area.
2. Biomass and carrying capacity of fodder/forage species along with prevailing population of livestock.
3. Experimentation on reported recipes and medicinal use for the evaluation of their toxicity and valuability for drug development.

# REFERENCES


Abdulla, P. 1973. In: Nasir, E. and S.I. Ali (eds.): *Flora of pakistan,* Sapindaceae:No. 1

Ahmad, M., A. Khan, M. Zafar and S. Sultan. 2006. Ethnobotanical Demography and Ecology Diversification of some important weeds from district Attock Pakistan. *Pak. J. Weed Sci. Res.,* 12(1-2): 37-46.

Ahmed, K., Z.I .Khan, M. Ashraf, M. Ibrahim and E.E. Valeem. 2008. Study of plant Diversity at Kufri (Soune Valley) Punjab, Pakistan and prevailing threats therein. *Pak. J. Bot.,* (3): 993-997.

Albert, L., A.L. Sajeum, R. Jayashree and N. Minaram. 2008. *Traditional tribal knowledge and Status of some rare and endemic medicinal plants of North Cachar Hills District Assam,* North East India.

Ali, S. I. and Y. J. Nasir. (Eds.). 1989-1991. *Flora of Pakistan* (Fascicle series). Islamabad, Karachi.

Ali, S.I. 1977. In: Nasir, E. and S.I. Ali (eds.): *Flora of Pakistan,* "Papilionacceae", 1pp.

Ali, S.I. and M. Qaiser. (Eds.). 1993-1995 & 2000-2008. *Flora of Pakistan* (Fascicle series) Islamabad, Karachi.

Anitha, B., V.R. Mohan, T. Athiperumalsami. 2008. ethnomedicinal plants used by the Kanikkars of Tirunelveli District, Tamil Nadu. *India to Treat Skin Diseases.* Ethnopharmacology unit, research department of Botany, V. O. Chidambaram College, Tuticorin, Tamil Nadu, India. Ethnobotany. (http://en.wikipedia.org/wiki/ethnobotany)

Anonymous, 2008a. *Ethnobotany* (http://en.wikipedia.org/wiki/ethnobotany)

Anonymous, 2008b. *Ethnobotany in Pakistan.* (http://www.wwfpak.org/ethno_pakistan).

Anonymous, 2008c. *District Cences Report,* Federal Beuro of Statics, Govt. of Pakistan.

Anonymous, 2009a. (http://www.encyclopedia.com/topic/National_parks.aspx)

Anonymous, 2009b. *Medicinal Plants of Nepal.* (http://www.pfaf.org/user/Plant.aspx?LatinName=Hedera+nepalensis)

Anonymous, 2012. (http://www.livestrong.com/article/479186-medicinal-benefits-of-malva-neglecta/#ixzz1woSgyEd7)

Anonymous. 1987. *Natural Food Institute,* Wonder Crops.

Arshad, M., Akbar and S. Rshid. 2002. *Wealth of Medicinal Plants of Chlistan Desert, Pakistan.* Hamdard Medicus, Vol. VI: 25-34.

Arshad, M., and A. Mushtaq. 2005. *Ethanobotanical study of Gilliyat for botanical Demography and Bio-ecological diversitfication.* Department of Botany University of Arid Agriculture Rawalpindi, Pakistan. Ethnobotanical leaflets. (http://www.siu.edu/phyto.htm)

Ashfaq, S., M. Ahmed and M.Arshad. 2004. Ethnobotanical observations of medically important plants of Tehsil Fateh Jang, Attock, Pakistan. *Pak. J. Bot.,* 7(1):25-33.

Asolkar, L.V., K.K. Kakkar and O.J. Chakre. 1992. *Second Supplement to Glossary of Indian Medicinal Plants with active principles.* Part-1 (A-K), CSIR, New Delhi.

Baquar, S.R. and M. Tasnif. 1984. Medicinal plants of southern West Pakistan. *Periodical Expert Book Agency,* D-42, Vivek Vehar, Delhi.

Baquar, S.R. and M. Tasnif. 1984. *Medicinal Plants of Southern West Pakistan, Periodical Expert Book Agency,* D-42, Vivek Vihar, Delhi-113334.

Baquar, S.R.H. 1989. *Medicinal and Poisonous Plants of Pakistan.* Pritas Karachi, Pakistan, 506p.

Bean, W. 1981.*Trees and Shrubs Hardy in Great Britain.* Supplement. Murray: Vol 1 - 4.

Bhatti, G.R., R. Qureshi and M. Shah. 2001. Ethnobotany of Qadan Wari of Nara Desert. *Pak. J. Bot.,* 33 (special issue), 801-812.

Bhatti, G.R., R. Qureshi and M. Shah. 2001. Ethnobotany of Qadan Wari of Nara Desert. *Pak. J. Bot.,* 33 (special issue), 801-812.

Bown, D. 1995. *Encyclopaedia of Herbs and their Uses.* Dorling Kindersley, London.

Chevallier, A. 1996. *The Encyclopedia of Medicinal Plants.* Dorling Kindersley, London.

Chiej, R. 1984. *Encyclopaedia of Medicinal Plants.* MacDonald.

Chopra, R.N., S.L. Nayar and I.C. Chopra. 1986. Glossary of Indian Medicinal Plants (Including the Supplement). *Council of Scientific and Industrial Research,* New Delhi.

Chudary, M.I., Q. He., Y.Y. Cheng and P.G. Xiao. 2006. Ethnobotany of medicinal plants from Tian Mu Shan biosphere reserve. Zhejiag-Province, China. *Asian J. Pl. Sci.,* 5(4): 646-653.

Coffey, T. 1993. *The History and Folklore of North American Wild Flowers.* Facts on File.

CSIR 2003. *The Wealth of India: a dictionary of Indian Raw Material and Industrial Products.* Raw material Vol. I-XI Revised Edition. Publications and Information Directorate, Council of Scientific and Industrial Research (CSIR), New Delhi, India.

Dobhal, U., B. Snehlata and N.S. Bisht. 2006. *Some medicinal weeds associated with terraces of crop fields of Pauri*, India. Department of Botany, H. N. B. Garhwal University Pauri (Garhwal), Uttaranchal. Ecology Diversification of some important weeds from district Attock Pakistan.

Duke, J.A. and E.S. Ayensu. 1985. *Medicinal Plants of China.* Reference Publications, Inc.

Duke, J.A. and Wain, K.K. 1981. *Medicinal plants of the world.* Computer index with more than 85,000 entries, vols 3.

Emboden, W. 1979. *Narcotic Plants,* Studio Vista.

Farnsworth, N.R and D.D. Soejarto 1991. Global importance of medicinal plants. In: Akerele, O., V. Heywood and H. Synge (Eds), conservation of medicinal plants.Cambridge University Press. *Fitoterapia* 79:132-141.

Foster, S. and Duke, J.A. 1990. *A Field Guide to Medicinal Plants. Eastern and Central N. America.* Houghton Mifflin Co.

Gamble, J. S. 1972. *A Manual of Indian Timbers.* Bishen Singh Mahendra Pal Singh.

Genders. R. 2009. *Scented Flora of the World.* (http://www.pfaf.org/user/Plant.aspx)

Ghazanfar, S. 1977. *Saxifragaceae.* In: Flora of Pakistan. (Eds.): E. Nasir & S.I. Ali. 108: 1-29, Islamabad.

Goodman, S.M., and A. Ghafoor. 1992. *The ethnobotany of Sourthern Balochistan, Pakistan with particular reference to Medicinal plants.* Fieldiana Bot., 31:1-84.

Grieve, 2009. A Modern Herbal (http://www.pfaf.org/user/Plant.aspx)

Grieve, M. 1984. *A Modern Herbal.* Penguin.

Grohmann, F. 1974. Oleaceae. In*: Flora of Pakistan* No. 59. (Eds.): E. Nasir and S.I Ali. Department of Botany Gordon College Rawalpindi. 1-27

Gupta, B.L. 1945. *Forest Flora of Chakrata,* Dehra Dun and Saharanpur. Forest Research Institute Press.

Haq, I.U. and M. Hussain. 1993. *Medicinal plants of Mansehra District.* Hamdard Medicus, 34: 63-99.

Harshberger, J.W. 1896. *The purpose of ethnobotany.* Botanical Gazette 21: 146-154

Hassan, V.U., M. Saleem, N. Shaffi, K. U. Din, M. Qasier and Asfandyar 2007. Lemon grass: Botany, Ethnobotany and Chemistry Review. *Pak J. Weed sci. Res.,* 13(1-2): 129-134.

Hazrat, A., J. Shah, M. Ali and I. Iqbal. 2007. Medicinal uses of Ranunculaceae of Dir Valley. *Pak. J. Bot.,* 39(4): 1037-1044.

Hedge, I. 1990. In: Nasir, E. and S.I. Ali (eds.): *Flora of Pakistan,* Labiatae, 192p.

Hill, A.F. 1952. *Economic Botany.* The Maple Press.

Launert, E. 1981. *Edible and Medicinal Plants.* Hamlyn.

Lust, J. 1983. *The Herb Book.* Bantam books.

Hussain, K., S. Amir and Z.H. Syed. 2008. *An ethnobotanical survey of important wild medicinal plants of Hattar Haripur, Pakistan.* Shakarganj sugar research institute, Toba Road Jhang (Punjab), Pakistan. Ethnobotanical leaflets. (http://www.siu.edu/phyto.htm)

Hussain, S. and F. Hassan. 2006. *Ethnobotanical evaluation of some plant resources in northern part of Pakistan.* Department of Botany G.MP. Graduate Jahanzeb College Swat, Pakistan. Depatment of Botany, University of Peshawar, Pakistan. Ethnobotanical leaflets (http:// www.siu.edu/phyto.htm)

Huxley A. 1992. *The New RHS Dictionary of Gardening.* MacMillan Press. ISBN 0-333-47494-5.

Idu, M., G.O. Obaruyi and J.O. Erhabor. 2009. *Ethnobotanical Uses of Plants Among the Binis in the Treatment of Ophthalmic and ENT* (Ear, Nose and Throat) Ailments. Ethnobotanical Leaflets 13: 480-96. (www.ethnobotanical leaflets.com)

Ikram, A. and Hussain M. 1978. Traditional use of herbs to get medicines, species in Himalayan Ranges. *Pak. J. Bot.,* 10(6): 65-74.

Ishtiaque, C.M., M.A. Khan and W. Hanif. 2006. An ethnomedicinal inventory of plants used for family planning and sex diseases treatment in Samahni valley, (A.K) Pakistan. *Pak. J. Biol. Sci.,* 9(14): 2546-2555.

Islam, M., A. Habib, R. Abdur, A. Naveed and K. Ikramullah. 2006. Weed and medicinal plants of Shawar Valley, District Swat. *Pak. J. Weed Sci. Res.,* 12(1-2); 83-88.

Jafri, S.M.H., 1966. *Flora of Karachi (Coastal West Pakistan) Book corporation Karachi* 1-4.

Jain, S.C., R. Jain and R. Singh. 2009. Ethnobotanical Survey of Sariska and Siliserh Regions from Alwar District of Rajasthan, India. *Ethnobotanical Leaflets* 13: 171-88. 2009. www.ethnobotanical leaflets.com.

Jeruto, P., C. Lukhoba, G. Ouma, D. Otieno and C. Mutai. 2008. An ethnobotanical study of medicinal plants used by the Nandi people in Kenya. *J. Ethnopharm.*116: 370-376.

Khan, S.W. and S. Khatoon. 2004. Ethnobotanical studies in Haramosh and Bugrote Valleys (Gilgit). *Int. J. Biotech.,* 1(4): 584-589.

Kirtikar, K. R. and B. D.Basu. 1933. *Indian Medicinal Plants partsIII,* Indian press Allahbad, india.

Koche, M.S., M. Saleem, N. Shaffi, K.U. Din, M. Qasier and Asfandyar. 2008. Lemon grass: Botany, E thnobotany and Chemistry Review. *Pak J. Weed Sci. Res.,* 13(1-2): 129-134.

Lentini, F. 2000. The role of ethnobotanics in scientific research; State of Ethnobotanical knowledge in Sicily. *Fitoterapia.* 71: 583 – 588.

Manandhar, N.P. 2002. *Plants and People of Nepal.* Timber. Press. Oregon.

Manandhar, N.P., 2009. *Plants and People of Nepal* (http://www.pfaf.org/user/ cmspage.aspx?pageid=187)

Martin, A. 1995. *Ethnobotany: A peoples and plants conscuation, amnual chapman and Hall.* London New, Tokyo.

Martin, A., M.A. Khan, M. Ashraf and R.A. Qureshi. 2007. Traditional use of herbs, shrubs and tree of Shogran Valley, Mansehra, *Pak. J. Boil. Sci.,* 4(9): 1101-1107.

Moerman, D. 1998. Native American Ethnobotany Timber Press. Oregon.

Mohan, V.R., A. Rajesh, T. Athiperumalsami and A. Sutha. 2008. Ethnomedicinal plants of the Tirunelveli district, Tamil Nadu, India. Ethnopharmacology unit, Research Department of Botany, V. O. Chidambaram College, Taticorin- 628008 Tamil Nadu, India. *Ethnobotanical Leaflets* 12: 79-95. (http://www.siu.edu/phyto.htm)

Nadkarni K. M. 1970. *Indian Material Medica.* Popular prakshan, Bombay, India.

Nasir, E. and S.I. Ali., 1970-1989. *Flora of Pakistan (fascicles series).* Department of Botany, University of Karachi.

Nasir, Y. J. 1983. In: Nasir, E. and S.I. Ali (eds.): *Flora of Pakistan,* Solanaceae:No. 168.

Nasir, Y.J. and Rafiq, R. 1995. *Wild Flowers of Pakistan.* Oxford University Press, Pakistan.

Niebuhr, A. D. 1970. *Herbs of Greece.* Herb Society of America.

NIH (National Institute of Health) 2003. *Monographs of Unani Medicine vol. I. Drugs Control and Traditional Medicine Division, National Institute of Health Islmabad Pakistan.* Asif Printing Agency, Rawalpindi, Pakistan. 664 pp.

Norscia, I.S. and M.B. Tarli. 2006. Ethnobotanical reputation of plant species from two forest of Madagagascar: A preliminary investigation. *South African J. Bot.,* 72: 656-660.

Oryem-Origa, H, E.K.Z. Kakudidi, A.B. Atende and Z.R. Bukenyan. 1995. *Preliminary ethnobotanical studies of the Rwenzon Mountain Forest Area*, Ugenda, Bothalia 25(1): 111-119.

Ozbuck, T.B., G. Hamd and E. Oznur. 2006. The contribution of wild edible plants to human nutrition in black Sea Region of Turkey. *Ethnobotany leaflets* 10: 98-103.

Parmar, C. and M.K. Kaushal. 1982. *Wild Fruits of the Sub-Himalayan Region.* Kalyani Publishers, New Delhi.

Passalacqua, N.G., P.M. Gurrera and G.D. Fine. 2007. Contribution to the knowledge of the folk plant medicine in Calabria Region. *Filtoterepia* 78: 52-68.

Pattantaik, C., C. Reddy and M.S.R. Murthy. 2008. An ethnobotanical Survey of medicinal plants used by the Didayi tribe of Malkangiri district of Orissa, India. *Fitoterapia* 79: 67-71.

Phillips, R. and M. Rix. 1991. *Perennials Volumes 1 and 2.* Pan Books.

Phillips. R. & Foy. N., 2009. Herbs (http://www.pfaf.org/user/Plant.aspx)

Polunin, O. & Stainton, A. 1987. *Concise flowers of the Himalaya.*

Polunin, O. and A. Huxley. 1987. *Flowers of the Mediterranean.* Hogarth Press.

Prasad, A., R.E.R. Zepeda and D.E. Johnson. 2008. Plant folk medicines and tree of Shogran Valley, Mansehra, *Pak. J. Boil. Sci.,* 4(9): 1101-1107.

Qaiser, 1978. Family Bombacaceae. In*: Flora of Pakistan,* Karachi University, Karachi, Pakistan, 119: 1-36.

Qureshi, R. 2002. Ethnobotany of Rohri hills, Sindh, Pakistan. *Hamdard Medicus*, 50(1): 156-162.

Qureshi, R. 2004. *Floristic and Ethnobotanical study of Desert Nara Region, Sindh.* Ph.D. Thesis, Department of Botany, Shah Abdul Latif University, Khairpur, Sindh, Pakistan 1: 1-30.

Qureshi, R. and G.R. Bahtti. 2007. *Wild Gourd: A green medicine, Hamdard Medicine* 50(1): 156-162.

Qureshi, R. and G.R. Bhatti. 2006. Ethnobotanical Observations of Achyranthes aspera L. And Aerva spp. with Special Reference to the People of Nara Desert. *Hamdard Medicus,* XLIX (1): 43-48.

Qureshi, R. and G.R. Bhatti. 2008. Ethnobotany of plants used by the Thari people of Nara Desert, Pakistan. *Fitoterapia* 79: 468–473

Qureshi, R., G.R. Bhatti and G.S. Jakhar. 2006. Taxonomy & Ethnobotany of Date palm in District Khairpur. *Hamdard Medicus,* XLIX (2): 121-125.

Qureshi, R., G.R. Bhatti and G.S. Jakhar. 2006. Taxonomy and ethnobotany of date palm in district Khairpur. *Hamdard and Medicus* XLIX (2): 121-126.

Qureshi, R., G.R. Bhatti and M. Shah. 2001. Ethnobotanical properties of *Aloe barbadensis* Mill. with particular reference to the peoples of Nara Desert. Hamdard Medicus, *XLIV* (3): 46-50.

Rafique, A., N. Ahmad and R.N. Malik. 2004. Ethnobotanical resources of district Sailkot. *Proceedings of international symposium on medicinal plants held at NARC,* 7-9 September Islamabad: 22.

Ram, J. 2004. *Assessment of habitat diversity, forest vegetation and human dependence in the buffer zone of Nanda Devi Biosphere of west Himalayas* 22-26.

Ramana, Y.V.V, C.S. Reddy and J.O. Erhabor. 2005. *Ethno- medico Botany of Maredumilli tribal region in the Eastern Ghats of Andhra Pradesh.*

Ramana, Y.V.V. and C.S. Reddy. 2001. *Ethno- medico Botany of Maredumilli tribal region in the Eastern Ghats of Andhra Pradesh.*

Rasool, G. 1998. Status of the flora land plant convention in the Sultanat of Oman. *Bio. Cornser* 85: 287-195.

Ratnam, F. and I. Raju. 2008. An ethnobotanical study of medicinal plants used by the Nandi people in Kenya. *J. Ethnopharm* 116: 370-376.

Reddy, G. M., P. Prashantkuma and D.E. Johnson. 2005. Traditional herbal remedies for gynecological disorders in women of Bidar district, Karanatakal, *India fitoterapia* 78: 48-51.

Rout, A. and G.A. Thatoi. 2009. Ethnobotanical studies of some important ferns. *Ethnobotanical Leaflets* 11: 164-172. http://www.siu.edu/ebl/ leaflets

Salazar, S.F.M., R.E.R. Zepeda and D.E. Johnson. 2008. Plant folk medicines for gastrointestinal disorders amoung the main tribes of Sonora, Mexico. *Fitoterapia* 79: 132-141.

Saroya, S. 2011. Herbalism, Phytochemistry and Ethnopharmacology *Amritpal Science Publishers,* 390–394.

Satti, S., 2000. *A hand book of Kotli Sattian. Kohsar writer forum Kotli Sattain.*

Shah, G.M. and M.A. Khan. 2006. *Common medicinal folk recipes of Siran Valley,* Mansehra, Pakistan. Ethnobotany leaflets.

Sharma, E., N. Shaffi, K.U. Din, M. Qasier and Asfandyar. 2004. The percentage of medicinal wild species in Himalayan Ranges. *Pak. J. Bot.,* 10(6): 65-74.

Sheikh, M.I. 1993. *Trees of Pakistan.* Winrock International Institute for Agricultural.

Shinwari, M.I. and M.A. Khan. 1998. *Ethnobotany of Margalla Hills National Park of Islamabad.* Department of Biological Sciences, Quaid-i-Azam University, Islamabad, Pakistan. PASTIC Printing Press. 1-100.

Shinwari, M.I., M.I. Shinwari and M. Shah. 2007. *Medicinal plants of Margalla Hills National Park,* Islamabad. Higher Education Commission, Islamabad.

Shinwari, M.L. and M.A. Khan. 1999. Ethnobotanical conservation status of Margalla hills National Park, Islamabad. *J. P. Research* 82: 53-60.

Singh, G. and P. Kachroo. 1976. *Forest Flora of Srinagar.* Bishen Singh Mahendra Pal Singh

Smith, A.R. 1986. In: Nasir, E. and S.I. Ali (eds.): *Flora of Pakistan, Euphorbiaceae* No. 172.

Srivastava, K. 2007. *Ethnobotanical studies of some important ferns. Ethnobotanical Leaflets* 11: 164-172. http://www.siu.edu/ebl/leaflets.

Srivastava, R.C. 1989. *Drug plant resources of Central India, Today and Tommorrow printers & Publishers,* 24 B/5, Desh Badhu, Gupta Road, Karolbagh-New Delhi.

Stewart, R.R. 1972. An Annotated Catalogue of Vascular plants from West Pakistan and Kashmir. In: E. Nasir, and S. I. Ali (eds.): *Flora of West Pakistan:* 74-102. Karachi.

Triska, D. 1975. *Hamlyn Encyclopaedia of Plants.* Hamlyn.

Tsarong and J.Tsewang. 1994. *Tibetan Medicinal Plants.* Tibetan Medical Publications, India.

Ugurlu, E. and O. Secmen. 2008. Medicinal plants popularly used in the village of Yunt Mountain (Manisa-Turkey) *Fiterapia* 79: 126-131.

Uphof, J.C.T. 1959. *Dictionary of Economic Plants.* Weinheim

Usher, G. 1974. *A Dictionary of Plants Used by Man.* Constable.

Usmanghani, K., A. Saeed and M.T. Alam. 1997. *Indusyunic medicne. Department of Pharmacognosy,* Faculty of Pharmacy, University of Karachi, Pakistan.

Vidyasagar, G.M. and P. Prashantkumar. 2007. Traditional herbal remedies for gynecological disorders in women of Bidar district, Karanatakal, India. *Fitoterapia* 78: 48-51.

Weiner, M.A. 1980. *Earth Medicine, Earth Food.* Ballantine Books.

Yeung, Him-Che. 1985. *Handbook of Chinese Herbs and Formulas.* Institute of Chinese Medicine, Los Angeles.

# APPENDICES

## APPENDIX 1: INDEX TO LOCAL NAMES

**(Continued)**

| | **Local name** | **Page No.** |
|---|---|---|
| 60 | Gao zeban | 31 |
| 61 | Garanda | 25 |
| 62 | Gidar buti | 25 |
| 63 | Gokhru | 31 |
| 64 | Gunger | 27, 30, 111 |
| 65 | Haleon | 28 |
| 66 | Harmal | 29, 52 |
| 67 | Harsingar | 28, 62, 106 |
| 68 | Hiran khuri | 25 |
| 69 | Isabgol | 29 |
| 70 | Itsit | 25, 31, 45 |
| 71 | Jal dhania | 30, 49 |
| 72 | Jal nim | 24 |
| 73 | Jangali matar | 28 |
| 74 | Jangli haloon | 26 |
| 75 | Jangli isabgol | 29 |
| 76 | Jangli Jai | 24 |
| 77 | Jangli mirch | 31 |
| 78 | Jangli palak | 30, 48 |
| 79 | Jawar | 31 |
| 80 | Jhari | 32 |
| 81 | Jungli podina | 28, 42 |
| 82 | Kach mach | 31 |
| 83 | Kahu | 29, 62, 106 |
| 84 | Kakoh | 27 |
| 85 | Kali jarri | 30, 43 |
| 86 | Kali jiri | 30 |
| 87 | Kameela | 28 |
| 88 | Kamlai | 27 |
| 89 | Kan phuti | 25, 34 |
| 90 | Kana | 30, 35, 107 |
| 91 | Kanair | 28, 56, 100 |
| 92 | Kandiali | 31, 66 |
| 93 | Kandiari | 31, 39 |
| 94 | Kangar | 29, 69, 100 |
| 95 | Kanikar | 29, 49 |
| 96 | Kanjura | 25 |

**(Continued)**

| | Local name | Page No. |
|---|---|---|
| 97 | Kauri buti | 24 |
| 98 | Khabbal ghaa | 26 |
| 99 | Khati buti | 29, 45 |
| 100 | Khatimber/Chuki | 30, 63 |
| 101 | Khokhal | 28, 61, 106 |
| 102 | Khub kalan | 31 |
| 103 | Kikri | 24 |
| 104 | Kusum | 26 |
| 105 | Leh | 25 |
| 106 | Leli | 25 |
| 107 | Loonder | 31, 108 |
| 108 | Lundur | 24 |
| 109 | Lunji | 31, 108 |
| 110 | Mudar | 25 |
| 111 | Mahori | 31, 65 |
| 112 | Maina | 28 |
| 113 | Majeth | 30 |
| 114 | Maraho | 30 |
| 115 | Marand | 27 |
| 116 | Matri | 31 |
| 117 | Murgha ghaa | 26 |
| 118 | Naghphani | 29, 58 |
| 119 | Nara | 18, 24, 138, 143 |
| 120 | Pahari poodina | 28 |
| 121 | Pahari tambako | 31 |
| 122 | Paleet | 25 |
| 123 | Panch phul | 28, 111 |
| 124 | Papra | 27 |
| 125 | Patakhi | 28, 102 |
| 126 | Peelan | 31, 50, 65 |
| 127 | Peepal | 27 |
| 128 | Persiaon shan | 24, 36 |
| 129 | Phagwara | 27, 105 |
| 130 | Phulai | 24, 104 |
| 131 | Piazi | 24 |

| | Local name | Page No. |
|---|---|---|
| 132 | Pirya Halim | 28 |
| 133 | Pohli | 25 |
| 134 | Puth kanda | 24, 56 |
| 135 | Ral | 28, 60, 104 |
| 136 | Rawari | 32, 34 |
| 137 | Rein | 30, 103 |
| 138 | Sadbarga | 31, 39, 101 |
| 139 | Samp di boti | 30 |
| 140 | Sanatha | 26 |
| 141 | Sathi | 30, 43 |
| 142 | Saunchal | 28 |
| 143 | Sedge Muther | 26 |
| 144 | Shahtra | 27 |
| 145 | Shakar Danna | 25 |
| 146 | Sheer bar | 26 |
| 147 | Shepherd's Purse | 25 |
| 148 | Shrinh | 24 |
| 149 | Siali | 29, 34 |
| 150 | Simbal | 25, 70, 102 |
| 151 | Sinji | 28 |
| 152 | Somni | 26 |
| 153 | Sonf | 27 |
| 154 | Sufeida | 29 |
| 155 | Sukhchain | 29 |
| 156 | Sumbul | 25 |
| 157 | Suranjan | 25, 44 |
| 158 | Taman | 27, 77, 110 |
| 159 | Tar rara | 26 |
| 160 | Tara mira | 26 |
| 161 | Tavi | 32, 68, 111 |
| 162 | Timbar | 32, 109 |
| 163 | Torki | 27 |
| 164 | Tut | 28, 61, 105 |
| 165 | Unt katara | 26 |
| 166 | Wilayati ak | 27, 33 |

# APPENDIX 2.
# INDEX TO BOTANICAL NAMES

| | ***Botranical name*** | **Page No** |
|---|---|---|
| 31 | *Bombax ceiba* L. | 25, 70, 78, 84, 102, 113, 119 |
| 32 | *Calotropis procera* (Willd.) R.Br. | 25, 84, 120 |
| 33 | *Cannabis sativa* L. | 5, 25, 84, 121 |
| 34 | *Capsella bursa-pastoris* (L.) Medik | 25, 84, 124 |
| 35 | *Cardiospermum halicacabum* L. | 5, 25, 34, 84, 121 |
| 36 | *Carissa opaca* Stapf ex. Haines | 25, 56, 78, 84, 100, 113, 119 |
| 37 | *Carthamus oxycantha* M. Bieb | 25, 84, 101, 121 |
| 38 | *Cenchrus ciliaris* L. | 25, 84, 124 |
| 39 | *Cenchrus pennisetiformis* Hochst. & Steud. | 25, 84, 124 |
| 40 | *Cenchrus setigerus* Vahl | 25, 84, 124 |
| 41 | *Chenopodium album* L. | 25, 41, 84, 120 |
| 42 | *Chenopodium ambrosioides* L. | 25, 84, 120 |
| 43 | *Chrysopogon aucheri* (Boiss.) Stapf. | 25, 85, 121 |
| 44 | *Cirsium arvense* (Linn.) Scope. | 24, 85.124 |
| 45 | *Clematis montana* Buch. | 5, 25, 85, 121 |
| 46 | *Clematis napaulensis* Royle | 5, 25, 85, 124 |
| 47 | *Colchicum aitchisonii* (Hook. f.) E. Nasir | 5, 25, 44, 78, 85, 121 |
| 48 | *Colebrookia oppositifolia* Sm. | 25, 59, 78, 85, 104, 120 |
| 49 | *Commelina paludosa* Blume | 25, 84, 124 |
| 50 | *Coniogramme rosthornii* Hieron. | 25, 85, 121 |
| 51 | *Convolvulus arvensis* L. | 5, 25, 85, 121 |
| 52 | *Conyza aegyptica* Ait. | 25, 85, 124 |
| 53 | *Conyza bonariensis* L. | 25, 85, 124 |
| 54 | *Conyza canadensis* L. | 25, 85, 124 |
| 55 | *Coronopus didymus* (L.) Sm. | 26, 85, 124 |
| 56 | *Crotolaria medicagnea* Lam. | 26, 85, 124 |
| 57 | *Cuscuta reflexa* Roxb. | 26, 54, 85, 121 |
| 58 | *Cynodon dactylon* L. Pers | 26, 86, 122 |
| 59 | *Cyperus rotundus* L. | 26, 54, 78, 86, 122 |
| 60 | *Dactyloctenium aegyptium* L. | 26, 86, 124 |
| 61 | *Datura stramonium* L. | 26, 59, 78, 86, 110, 120 |
| 62 | *Desmostachya bipinnata* (L.) Stapf | 26, 86, 122 |
| 63 | *Dicanthium annulatum* (Forssk.) Stapf. | 86, 107 |
| 64 | *Dicliptera roxburghiana* Nees | 5, 26, 86, 124 |

**(Continued)**

| | ***Botranical name*** | **Page No** |
|---|---|---|
| 65 | *Digera muricata* (L.) Mart. | 5, 26, 86, 124 |
| 66 | *Dioscorea deltoidea* Wall. ex Kunth | 26, 86, 122 |
| 67 | *Diospyros lotus* L. | 26, 70, 78, 86, 102, 113, 119 |
| 68 | *Dodonaea viscosa* (L.) Jacq. | 26, 65, 86, 110, 120 |
| 69 | *Echinochloa crus-galli* (L.) P. Beauv. | 26, 86, 113, 122 |
| 70 | *Echinops echinatus* Roxb. | 26, 87, 124 |
| 71 | *Eclipta prostrata* (L.) L. | 26, 87, 124 |
| 72 | *Equisetum ramosissimum* (Desf.) | 26, 124 |
| 73 | *Eragrostis ateroviens* (Desf.) Trin. ex Nees | 26, 87, 124 |
| 74 | *Eragrostis minor* Host. | 26, 87, 124 |
| 75 | *Eruca sativa* L. | 26, 40, 87, 101, 119 |
| 76 | *Euphorbia clarkeana* Hkf. | 26, 87, 124 |
| 77 | *Euphorbia granulata* Forssk. | 26, 87, 124 |
| 78 | *Euphorbia helioscopia* Mewski. | 26, 87, 125 |
| 79 | *Euphorbia hirta* L. | 26, 87, 125 |
| 80 | *Euphorbia indica* Lam. | 26, 87, 125 |
| 81 | *Euphorbia prostrata (*L.) Ait | 27, 87, 125 |
| 82 | *Euphrasia himalayica* Wettst. | 27, 87, 125 |
| 83 | *Ficus bengalensis* L. | 27, 74, 78, 87, 105, 120 |
| 84 | *Ficus carica* L. | 27, 60, 87, 105, 118 |
| 85 | *Ficus palmata* Forssk. | 27, 87, 105, 120 |
| 86 | *Ficus religiosa* L. | 27, 87, 119 |
| 87 | *Ficus roxburghii* Wall. ex Brand. | 27, 74, 78, 88, 105, 113, 119 |
| 88 | *Flacourtia indica* (Burm. f.) Merr. | 27, 88, 118 |
| 89 | *Foeniculum vulgare* (Miller) | 27, 88, 119 |
| 90 | *Fumaria indica* (Hausskn.) H.N. Pugsley | 27, 88, 125 |
| 91 | *Gallium aparine* L. | 5, 27, 88, 125 |
| 92 | *Geranium rotundifolium* L. | 5, 27, 88, 125 |
| 93 | *Grewia optiva* Drum. ex. Burret. | 27, 88, 119 |
| 94 | *Grewia tenax*(Forssk.) Aschers & Schweinf. | 27, 88, 125 |
| 95 | *Hedera nepallensis* K. Koch | 5, 27, 88, 120 |
| 96 | *Heliotopium europaeum* L. | 27, 88, 122 |
| 97 | *Imperata cylindrica* (L.) Raeuschel | 27, 88, 125 |

| | *Botranical name* | Page No |
|---|---|---|
| 98 | *Indigofera himalayensis* Ali | 27, 88, 103, 122 |
| 99 | *Indigofera linifolia* (L. f) Retz | 27, 89, 125 |
| 100 | *Indigofera sessiliflora* DC | 27, 89, 125 |
| 101 | *Ipomoea carnea* Jacq. | 27, 33, 78, 122 |
| 102 | *Ipomoea hederacea* (L.) Jacq. | 27, 89, 125 |
| 103 | *Ipomoea nil* (L.) Roth | 27, 89, 125 |
| 104 | *Kickxia ramosissima* (Wall.) Janchen | 27, 89, 125 |
| 105 | *Lactuca auriculata* (Wall. ex Dc.) | 27, 125 |
| 106 | *Lactuca dissecta* D. Don. | 27, 89, 125 |
| 107 | *Lactuca serriola* L. | 27, 89, 125 |
| 108 | *Lannea coromandelica* L. | 27, 122 |
| 109 | *Lantana camara* L. | 28, 111, 120 |
| 110 | *Lantana indica* Roxb. | 28, 111, 125 |
| 111 | *Lathyrus aphaca* L. | 5, 28, 89, 125 |
| 112 | *Launaea procumbens* (Roxb.) Ram. & Rajgo. | 28, 39, 79, 89, 125 |
| 113 | *Lepidium sativum* L. | 28, 89, 125 |
| 114 | *Lotus corniculatus* (Wald. & Kit. ex Willd.) Briq. & Rech. F. | 28, 125 |
| 115 | *Mallotus philipensis* (Lam.) Muell. | 28, 70, 79, 89, 102, 114, 11 |
| 116 | *Malva neglecta* Waller. | 28, 44, 120 |
| 117 | *Malvastrum coromendelianum* L. | 5, 28, 90, 122 |
| 118 | *Maytenus royleanus* (Wall. ex Lawson) Cufodontis | 28, 90, 102, 114, 120 |
| 119 | *Medicago denticulata* Willd. | 28, 90, 114, 122 |
| 120 | *Medicago laciniata* (L.) Mill. | 28, 90, 122 |
| 121 | *Medicago polymorpha* L. | 5, 28, 90, 122 |
| 122 | *Melia azedarach* L. | 28, 72, 104, 122 |
| 123 | *Melilotus indica* (L.) All. | 28, 90, 114, 125 |
| 124 | *Mentha longifolia* (L.) Huds | 5, 28, 90, 120 |
| 125 | *Mentha royleana* Wall. ex Bth. | 28, 42, 122 |
| 126 | *Micromeria biflora* (Ham.) Bth. | 5, 28, 90, 114, 122 |
| 127 | *Mimosa rubicaulis* Lam. | 28, 60, 90, 104, 122 |
| 128 | *Morus alba* L. | 28, 74, 90, 105, 118 |
| 129 | *Morus nigra* L. | 10, 28, 61, 79, 90, 105, 114, 118 |
| 130 | *Myrsine africana* L. | 28, 61, 90, 106, 119 |
| 131 | *Nasturtim officinale* R. Br. | 28, 114, 125 |

**(Continued)**

| | *Botranical name* | Page No |
|---|---|---|
| 132 | *Nerium oleander* L. | 28, 56, 100, 114, 122 |
| 133 | *Nyctanthes arbor-tristis* L. | 28, 62, 79, 90, 106, 114, 120 |
| 134 | *Olea europaea* L. | 28, 75, 79, 90, 106, 114, 119 |
| 135 | *Olea ferruginea* Royle | 4, 29, 90, 119 |
| 136 | *Opuntia monacantha* Haw. | 29, 58, 114, 122 |
| 137 | *Origanum vaulgare* L. | 29, 42, 79, 91, 114, 122 |
| 138 | *Otostegia limbata* (Benth.) Boiss. | 29, 59, 91, 122 |
| 139 | *Oxalis corniculata* L. | 29, 45, 91, 120 |
| 140 | *Parthenium hytserophorus* L. | 29, 91, 125 |
| 141 | *Peganum harmala* L. | 29, 52, 79, 91, 122 |
| 142 | *Periploca aphylla* Decne. | 29, 57, 79, 101, 115, 122 |
| 143 | *Phalaris minor* Retz. | 29, 91, 125 |
| 144 | *Phyla nodiflora* (L.) Greene | 29, 51, 79, 91, 122 |
| 145 | *Phyllanthus emblica* L. | 29, 102, 115, 119 |
| 146 | *Physalis divaricata* D. Don. | 50, 79 |
| 147 | *Physalis minima* L. | 29, 91, 115, 125 |
| 148 | *Pinus roxburghii* Sargent | 29, 76, 91, 106, 119 |
| 149 | *Pistacia chinensis* Bunge | 29, 69, 91, 100, 120 |
| 150 | *Plantago lanceolata* L. | 5, 29, 46, 91, 122 |
| 151 | *Plantago major* L. | 29, 46, 91, 122 |
| 152 | *Plantago ovata* Frossk. | 29, 79, 91, 122 |
| 153 | *Polygonum barbatum* L. | 29, 125 |
| 154 | *Polygonum plebejum* R. Br. | 29, 48, 79, 92, 115, 120 |
| 155 | *Polypogon fugax* Nees ex Steud. | 29, 92, 126 |
| 156 | *Pongamia pinnata*(L.) Pierre | 29, 92, 119 |
| 157 | *Populus deltoides* Bartram ex Marsh. | 29, 92, 109, 115, 123 |
| 158 | *Potentilla supina* L. | 29, 49, 92, 123 |
| 159 | *Psammogeton biternatum* Edgew. | 29, 92, 126 |
| 160 | *Pteridium aquilinum* (L.) Kuhn | 29, 92, 126 |
| 161 | *Pueraria tuberosa* (Roxb. ex Willd.) DC. | 29, 34, 79, 92, 123 |
| 162 | *Punica granatum* L. | 29, 63, 92, 108, 118 |

| | ***Botranical name*** | **Page No** |
|---|---|---|
| 163 | *Pyrus pashia* Ham. ex D. Don | 92, 109, 115 |
| 164 | *Quercus dilatata* Lindley | 30, 92, 120 |
| 165 | *Quercus incana* Roxb. | 30, 92, 103, 115, 120 |
| 166 | *Quercus leucotrichophora* | 4, 7, 30, 72, 79, 92, 103, 115, 119 |
| 167 | *Rananculus muricatus* L. | 30, 123 |
| 168 | *Rananculus sceleratus* L. | 5, 30, 93, 126 |
| 169 | *Ranunculus arvensis* L. | 30, 49, 123 |
| 170 | *Reinwardtia indica* Dumort. | 30, 44, 115, 123 |
| 171 | *Rhynchosia minima* (L.) DC | 5, 30, 93, 126 |
| 172 | *Ricinis communis* L. | 30, 123 |
| 173 | *Rosa brunonii* Lindll. | 30, 64, 79, 93, 109, 115, 119 |
| 174 | *Rubia cordifolia* L. | 5, 30, 34, 79, 93, 123 |
| 175 | *Rubus ellipticus* Smith | 30, 54, 93, 109, 119 |
| 176 | *Rumex dentatus* L. | 30, 48, 79, 93, 120 |
| 177 | *Rumex hastatus* D. Don. | 30, 93, 119 |
| 178 | *Rumex nepalensis* Spreng | 30, 48, 80, 93, 123 |
| 179 | *Saccharum bengalense* Retz. | 30, 93, 107, 116, 120 |
| 180 | *Saccharum spontaneum* L. | 30, 93, 107, 120 |
| 181 | *Sageretia thea* (Osback) M. C. Johnston | 30, 93, 108, 120 |
| 182 | *Salvia moorcroftiana* Wall. ex Bth. | 30, 43, 80, 93, 123 |
| 183 | *Salvia plebeia* R. Br. | 5, 30, 43, 93, 116, 120 |
| 184 | *Sauromatum venosum* (Ait.) Schott. | 30, 94, 123 |
| 185 | *Saussurea albescens* (DC.) Schr. Bip. | 30, 94, 126 |
| 186 | *Saussurea atkinsonii* (Clarke) | 30, 94, 126 |
| 187 | *Saussurea heteromalla* DC. | 30, 94, 126 |
| 188 | *Setaria glauca* (L.) P. Beauv | 116 |
| 189 | *Sida cordata* (Burm. f.) Borss.-Waalkes | 31, 44, 80, 94, 123 |
| 190 | *Silene conoidae* L. | 31, 94, 126 |
| 191 | *Silybum marianum* (L.) Gaertin | 31, 39, 94, 120 |
| 192 | *Sisymbrium irio* L. | 31, 40, 80, 94, 123 |
| 193 | *Smilax aspera* L. | 5, 31, 35, 80, 94, 123 |
| 194 | *Solanum incanum* L. | 65, 80, 94, 120 |
| 195 | *Solanum nigrum* L. | 31, 65, 94, 120 |
| 196 | *Solanum pseudocapsicum* L. | 31, 116, 126 |
| 197 | *Solanum surattense* Burm.f. | 31, 66, 94, 121 |
| 198 | *Solanum villosum* (L.) Moench | 31, 50, 94, 121 |

**(Continued)**

| | ***Botranical name*** | **Page No** |
|---|---|---|
| 199 | *Sonchus arvense* | 31, 94, 126 |
| 200 | *Sonchus asper* (L.) Hill. | 31, 94, 126 |
| 201 | *Sorghum bicolor* (L.) Moench. | 31, 95, 126 |
| 202 | *Sorghum halepense* (L.) Bern. | 31, 95, 121 |
| 203 | *Stellaria media* (L.) Cry. | 5, 31, 95, 123 |
| 204 | *Tagetes minuta* L. | 5, 31, 39, 80, 101, 116, 123 |
| 205 | *Taraxacum officinale* Weber. | 5, 31, 95, 121 |
| 206 | *Taraxcum wallichii* DC. | 31, 95, 126 |
| 207 | *Themeda anathera* (Nees) Hack. | 108, 117 |
| 208 | *Trianthema portulacastrum* L. | 31, 95, 126 |
| 209 | *Tribulus terrestris* L. | 31, 53, 80, 95, 123 |
| 210 | *Trichodesma indicum* (L.) R. Br. | 31, 40, 80, 95, 123 |
| 211 | *Trifolium repens* L. | 5, 31, 95, 126 |
| 212 | *Urtica dioica* L. | 31, 123 |
| 213 | *Valeriana wallichii* DC. | 31, 51, 80, 95, 123 |
| 214 | *Verbascum thapsus* L. | 31, 49, 95, 110, 119 |
| 215 | *Verbena officinalis* L. | 31, 51, 80, 95, 123 |
| 216 | *Vicia faba* L. | 5, 31, 95, 126 |
| 217 | *Vicia sativa* L. | 32, 34, 95, 123 |
| 218 | *Viola cancescens* Wall. ex Roxb. | 32, 80, 95, 123 |
| 219 | *Vitex negundo* L. | 32, 68, 96, 111, 117, 119 |
| 220 | *Withania somnifera* (L.) Dunal. | 32, 67, 96, 110, 121 |
| 221 | *Woodfordia fruticosa* (L.) S. Kurz | 5, 32, 96, 111, 121 |
| 222 | *Xanthium strumarium* L. | 32, 101, 126 |
| 223 | *Zanthoxylum alatum* Roxb. | 32, 80, 96, 117, 119 |
| 224 | *Zizyphus mauritiana* Mill. | 32, 76, 96, 108, 119 |
| 225 | *Zizyphus oxyphylla* Edgew. | 5, 32, 64, 96, 109, 119 |

# INDEX

## A



## B

## C



## D



## E

## I



## J



## K



## L



## M

## N



## O



## P



## Q



## R



## S

## T



## U



## V

## W



## Y